Active Debris Removal in Space

How to Clean the Earth's Environment from Space Debris

ROBIN BIESBROEK

PREFACE

Astronaut Wubbo Ockels of the European Space Agency (ESA) once said 'We are all astronauts of spaceship Earth'. It is a common phenomenon that humans realize the vulnerability of the Earth when looking at it from space, and that we should feel the responsibility to maintain the environment of our Earth in good health. For decades now the environmental focus has been dominated by global warming, the pollution of our oceans, and the destruction of rain forests.

However, the environment of the Earth extends beyond the atmosphere, The Earth's gravitational sphere of influence stretches to about 1 million kilometers, meaning that all Earth observation, science, communication and weather satellites, as well as our natural satellite the Moon, move in a space that is part of the environment of the Earth.

The fact is that this environment in space is also getting polluted; pieces of space debris such as old rocket upper stages, derelict satellites, lost repair tools and parts of exploded satellites make up 95% of all traceable objects in space. What is more important is that much of this space debris will stay in orbit for many years, even centuries.

Large pieces of space debris enter the Earth's atmosphere on a weekly basis, operational satellites have to make frequent Collision Avoidance Maneuvers, and the amount of debris in space keeps increasing.

It is time that we start to realize that our environment includes space around the Earth. Satellites in space provide us with many benefits, including: navigation directions for cars and planes, monitor pollution of the Earth, thickness of ice, and the weather. They help us predict the weather forecast, and help us to communicate with other continents. They even provide communication systems in the most remote places, provide us with hundreds of TV channels, provide us with new solutions within bio-science, and new technologies such as microchips and fuel cells. And finally, they give us stunning pictures of the Earth, Moon, Milky Way and even the entire universe which helps us understanding how the entire universe if formed. With the risk of space debris destroying satellites by collisions, these advantages are under a threat. In space, an impact by an object the size of a human hand with a speed of almost 50,000 km/hour (31,000 mph) leads to an explosion similar to that of a hand grenade. This problem can only be solved if space debris is removed from Earth's orbit by starting to actively remove space debris from orbit.

This book highlights the problems related to space debris, both technical (chapter 1) and legal (chapter 2), and shows options to start cleaning up space (chapter 3). Design aspects of a space garbage truck are shown in terms of grabbing debris (chapter 4) and creating a vehicle to transport it (chapter 5). Finally, some worries and dreams of the future are described in chapter 6.

My inspiration came from working with the Clean Space people at ESA, and therefore I would like to thank all involved in Clean Space!

A special thanks to Andrew Wolahan (e.deorbit system engineer at ESA) for giving me the idea to write this book and for giving very valuable feedback on the contents.

A big 'thank you' to Andrew Pickering (technical author at ESA's Concurrent Design Facility) for checking the spelling and wording.

Finally, I would like to thank Mónica Martínez Fernández (TT&C engineer for ESA's GAIA, SMART-1, ExoMars and Sentinels missions) for endless support and improving the way I present the topics described in this book.

CHAPTER 1: WHAT IS SPACE DEBRIS AND WHY IS IT A PROBLEM?

Before brainstorming about solutions to clean up space, let's first understand the gravity of the situation: the size of the debris cloud, where debris is located, what problems did it give in the past and what problems can we expect in the future.

HOW IT ALL STARTED

In the night of the 4th of October 1957 the Sputnik rocket blasted into space from its launch pad in Baikonur. Almost five minutes later, the Soviet Union had launched the world's first artificial satellite: Sputnik 1. The small satellite, less than 100 kg, was now orbiting the Earth every 1.5 hours and transmitting signals from space until its batteries ran out of power less than a month later.

While it was an enormous achievement in itself, the end of life of Sputnik 1 started a new era in space: an era of humans producing waste in space. Sputnik 1 had become the first space debris: an object made by humans, without function, and without possibility to recycle it or dispose of it. The rocket had put it into an elliptic orbit around the Earth: the lowest point (called perigee) 215 km high, and the highest point (called apogee) 947 km high. For space applications, an altitude of 215 km is low enough for the satellite to experience drag effects from the Earth's atmosphere. While this effect is small as the atmosphere at that altitude is incredibly thin, the effect was large enough to lower the orbit of Sputnik 1 for the following months. On 4 January 1958, it entered the dense regions of the atmosphere, disintegrated due to the enormous aero thermal forces, and burned up. The time Sputnik 1 spent in space was four months, of which it was active for less than one month, therefore spending most of its lifetime as space debris.

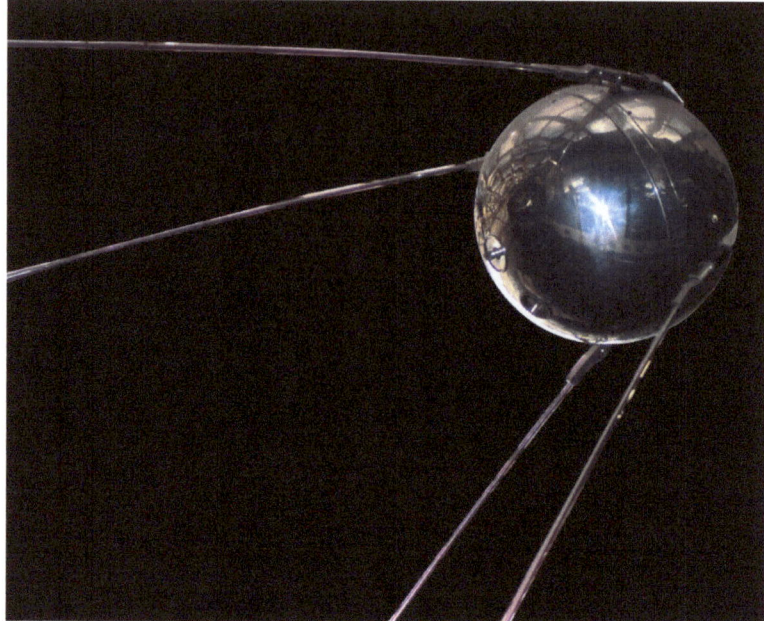

Figure 1: Sputnik 1. Credits: NASA.

However Sputnik 1 was not the only debris in orbit. When a rocket launches a satellite, several rocket stages are used. The final stage of the rocket has the task of delivering the satellite in the exact orbit that it requires. However this can only be achieved by achieving that orbit itself, and then gently releasing the satellite from the last rocket stage (also referred to as 'upper stage'). This means that when Sputnik 1 was delivered in orbit, the upper stage of the Sputnik rocket was also present in exactly the same orbit. The function of upper stages stops when the satellite is delivered, so from that point on an upper stage also qualifies as space debris. The lower stages of the rocket normally fall back to Earth. While the Sputnik 1 satellite was less than 100 kg and less than 1 meter diameter, the upper stage that delivered it into orbit was 7500 kg and 26 meters long!

Some people who looked up at the night sky in 1957 claimed to have seen Sputnik 1, but if they saw a moving dot in space it was actually the bigger upper stage that they saw.

During its launch on 4 October 1957, the satellite needed to be protected from the friction with the atmosphere. This was done by mounting a cone-type heat resistant structure over the satellite, attached to the rocket upper stage. This structure is called a fairing and is not required from the moment that space is reached, since friction in space is very low. The fairing was ejected from the rocket upper stage before the Sputnik 1 satellite was released, which means that yet another piece of space debris was resulting from this one launch. Since all these objects were delivered in similar orbits, all of them reentered the atmosphere within a few months. The problem of leaving trash in space had solved itself. However more was to come...

THE SITUATION NOW

Let's fast-forward to today. At the time of writing this book, there are roughly 17,000 objects in space tracked from Earth that are larger than a coffee cup. Only 600 of these objects are active satellites: less than 4%. All other objects qualify as space debris. Objects smaller than 5 to 10 cm in Low Earth Orbit (LEO) cannot be tracked from Earth; however computer models predict that over 750,000 objects larger than 1 cm exist.

Satellites in popular orbits now get collision warnings on a weekly basis. A collision warning is triggered when trajectory propagation models predict that the chance of the satellite colliding with another object is above a certain threshold. In this case, an analysis needs to be done with high precision trajectory propagation models to see if there is truly a chance of colliding, and if so, a Collision Avoidance Maneuver needs to be executed. A Collision Avoidance Maneuver, also known as CAM, lowers or raises the altitude of the satellite slightly in order to miss the object. Afterwards, the altitude is corrected again to the nominal altitude that the satellite needs to fulfill its mission. This CAM can only be done if the satellite has a propulsion system (small rockets on board the satellite) and costs fuel which is normally used to keep the mission functional. Apart from this, a CAM often interrupts the services that a satellite needs to give. Often the satellite needs to be turned into the right direction to give the maneuver, and this could imply that the sensors on board the satellite do not point anymore in the right direction (to Earth, for example). A CAM reduces the propellant on board, decreasing the operational lifetime and hence reducing the overall return of the initial investment. Finally, the satellite could have different modes of operations, and the mode to perform maneuvers is typically different from the mode to observe, and observation sensors may therefore have to be switched off when executing maneuvers.

Spacecraft that do not have a propulsion system to boost themselves up and down will need to hope for the best if there is a confirmed collision alert. Or in some cases, a warning may come too late. It has happened more than once that astronauts on board the International Space Station (ISS) needed to shelter in a Soyuz capsule, after a collision warning was issued. The ISS carries humans on board, and debris may actually cause a life threatening situation.

However another event is happening every week: the reentry of debris from space into the atmosphere of the Earth. Every week another object enters the atmosphere and normally burns up. However in some cases, typically for objects heavier than 1000 kg, parts of the satellite may survive the heated path through the atmosphere and fall down to Earth. Several parts of rocket stages have crashed on Earth in the last few years, such as a stainless steel fuel tank of a Delta 2 rocket that crashed near Georgetown, Texas on 22 January 1997, see Figure 2.

Figure 2:   Delta 2 space debris near Georgetown. Credits: NASA.

So what happened in between the Sputnik 1 launch and now? Let's have a look at how the number of objects in space is increasing, and why.

OVERVIEW OF SPACE DEBRIS

In order to get a global picture of space debris, we first need to understand what space debris exactly is. The Inter-Agency Space Debris Coordination Committee (IADC) defines space debris according to the following features:

- Objects including fragments and elements
- Made by humans
- In Earth orbit or reentering the atmosphere
- Non-functional

Little meteorites in space are therefore not considered space debris, as they were not made by us. Also any functioning satellite cannot be considered debris, but basically anything else can.

Figure 3 gives a perfect overview of space debris since the first Sputnik launch until 2013. It shows the evolution of objects in space that we can track from Earth.

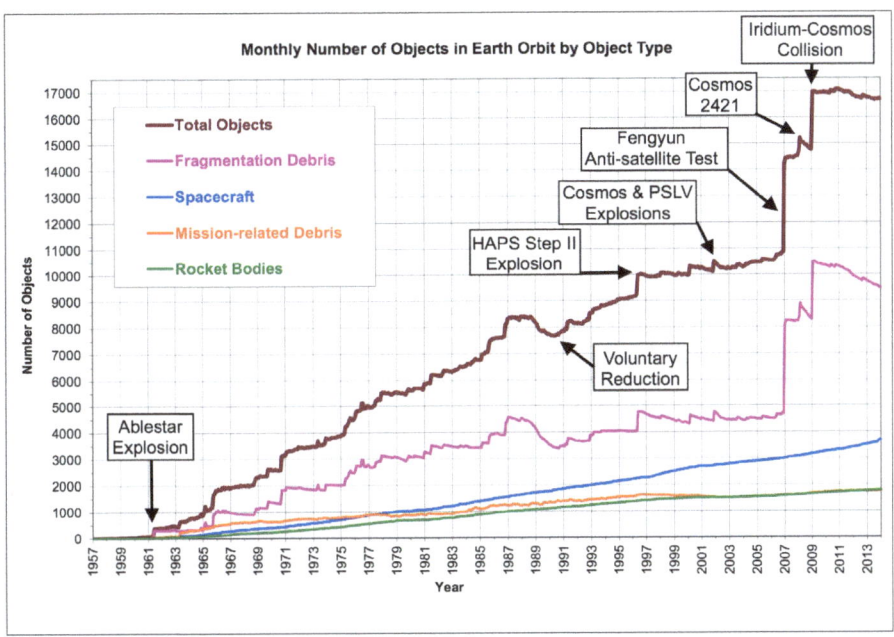

Figure 3: Evolution of the number of objects in Earth orbit. Credits: NASA, annotated by Mika McKinnon [RD1].

The top line indicates the total number of objects. We see that by the end of the 60s, there were already 2500 objects in space, and this number increased steadily by another 2500 every decade. Then after the millennium things went wrong. In one decade the number of objects increased from 10,000 to 17,000. Are all these satellites? No, looking at the blue line in the above plot, we see that the number of satellites increased linearly since Sputnik 1 to just over 3500 satellites today. Therefore, satellites represent only 20% of all objects in orbit around the Earth (that we can track). And today only about 600 satellites are functional.

Rocket bodies, represented by the green line at the bottom, are in minority but still represent a large total mass in orbit due to their large size and high mass. Similar to satellite bodies, their number in orbit increased almost linearly to almost 2000 today, therefore representing about 12% of all traceable objects in orbit.

So what are all other objects? The remaining 74% is classified as mission-related debris and fragmentation debris. Mission-related debris is debris parts that were created as part of a mission. For example, a cover is removed from a telescope and left in orbit. The number of mission-related objects is almost similar to the number of rocket bodies: around 2000.

Fragmentation debris is the biggest problem in terms of number of objects in Earth orbit: their number increased rapidly over the past few decades and now over 9000 fragments are tracked, three times the amount of satellites.

They represent more than half of all objects in orbit around the Earth. We can also see from the plot that there were occasions where the number of fragments, and therefore total amount of objects, decrease. For example in the 80s, when the first realization of space debris as a problem occurred and first actions for space debris remediation were taken. Also the current trend since 2011 is downhill, as the rate of fragments entering the atmosphere is higher than the rate of new objects in space. So how were these fragments created? By explosions and collisions.

EXPLOSIONS

It did not take very long for the first explosion in space to occur: four years after Sputnik 1, an American satellite Transit 4A was launched by a Thor-Ablestar rocket to an altitude of almost 1000 km. While this seemed like mission success, a bit more than an hour later the upper stage of the Thor-Ablestar rocket suddenly exploded. Almost 300 traceable fragments were created representing over 600 kg in total. Just before the explosion, the number of objects orbiting Earth was 50 (of which almost half were space debris). Afterwards, this number had increased to 350; of which now most of them were space debris. The incident can clearly be seen in Figure 3 where there is a steep increase in fragmentation debris in 1961. The cause of the explosion is unknown. Almost 40 years later, in 1998, 200 of those fragments were still being tracked from ground! This clearly shows a new problem: if debris is located at high enough altitudes, it may stay in orbit for decades, even centuries.

Since 1961, explosions have occurred on a yearly basis. Some are unintentional, but there were intended explosions too. In the early sixties, the Soviet Union would intentionally explode their satellites in order to prevent them from falling into the hands of the US, when the satellites reenter.

Another explosion worth noting is the HAPS explosion in 1996 at an altitude of 625 km. The HAPS (Hydrazine Auxiliary Propulsion System) was an upper stage of a Pegasus rocket, launched two years before. The number of fragments produced was larger than what computer models predicted at that time, and therefore these models needed to be updated. However the most important impact of the explosion is that the cause was determined after thorough analysis that there was still fuel present in the stage. This immediately created the need for a new procedure to be implemented in order to avoid future explosions: satellites or upper stages should deplete themselves of any fuel when their tasks have been completed. For upper stages, the depletion could also partially be used to lower the orbit of the upper stage, causing a quicker reentry into the atmosphere and thus helping to reduce space debris.

In a similar fashion, it was discovered that batteries can also explode in space if they get overcharged causing the spacecraft to break up, see Figure 4 below. The Ekran 2 explosion in 1978 is an example of this phenomenon.

Figure 4: Battery explosion. Credits: ESA.

This also led to a new rule within space agencies where batteries need to be depleted at the end of the satellite's life. Measures need to be taken to ensure that a satellite cannot come back to life and start charging the batteries. This can be done by wiping out the memory of the on board computer.

COLLISIONS

1996 was an important year for space debris. Not only did it see the explosion of the HAPS upper stage due to remaining fuel, it also saw the first witnessed natural collision in space. The CERISE spacecraft of the French government collided on 24 July with an Ariane rocket fragment. The 50 kg spacecraft, of British design, used a gravity gradient boom to stabilize itself. A 6 meter tall boom was deployed with a small mass on top. Due to the difference in gravity force between the mass on top of the boom, and mass of the satellite itself, the system stabilizes itself along a vertical axis pointed towards Earth. Spacecraft operators suddenly noticed a change in attitude, which could not be explained and was considered an anomaly. It the meantime The US Space Surveillance Network (SSN) noticed that a new object had appeared in space. After analysis, it was concluded that an Ariane 4 rocket body fragment had chopped off part of the gravity gradient boom (see Figure 5). Luckily therefore, only one fragment of debris was created. Calculations indicated that the impact occurred with a relative velocity of almost 15 km/second, meaning almost 54,000 km/hour (34,000 mph).

Figure 5: Artist impression of CERISE's gravity gradient boom being hit by space debris in 1996. Credits: CNES.

More analysis was done in the next years, and in 2005 it was discovered that an earlier collision had taken place in 1991, between Russian navigation satellite Cosmos 1934 and a piece of debris from a similar spacecraft, Cosmos 926. However before this first natural collision, many intended collisions had already taken place.

In 1970 a sudden increase can be seen in the debris overview graph Figure 3 which was a result of the Soviet anti-satellite program: Cosmos 374 was launched on October 23 and exploded into more than 100 traceable pieces, four hours after launch. Then, Cosmos 375 intercepts Cosmos 373 and explodes as well into more than 40 traceable pieces. About 20 of those tests were performed by the Soviet Union, producing more than 700 fragments. Many of these pieces are still in orbit today. Also the United States started showing that they could destroy their own satellites in the 80s, though with an attempt to not produce long-lasting debris in Earth orbit. These tests continued until by collective international agreement, anti-satellite systems producing space debris were banned. This ban was valid for 20 years, until on 11 January 2007 the Chinese launched an anti-satellite test targeting the Feng Yun-1C weather satellite. This successful test created the largest space debris cloud to date: over 3300 fragments were produced, increasing the total population of traceable space debris by 25%, as can be seen in Figure 3. Moreover, the debris fragments were not confined to the original orbit of Feng Yun-1C (about 850 km altitude); fragments were tracked up to 4000 km altitude, and as low as 200 km altitude. Satellites over a wide range of orbits could be in the path of fragments for many years to come. A true devastation in space. Figure 6 shows the different orbits in which fragments of Feng Yun-1C are scattered, one month after the collision. By now, fragments are scattered around all over the globe in low Earth orbits and it is expected that by the year 2100, 10% of those fragments will still orbit the Earth.

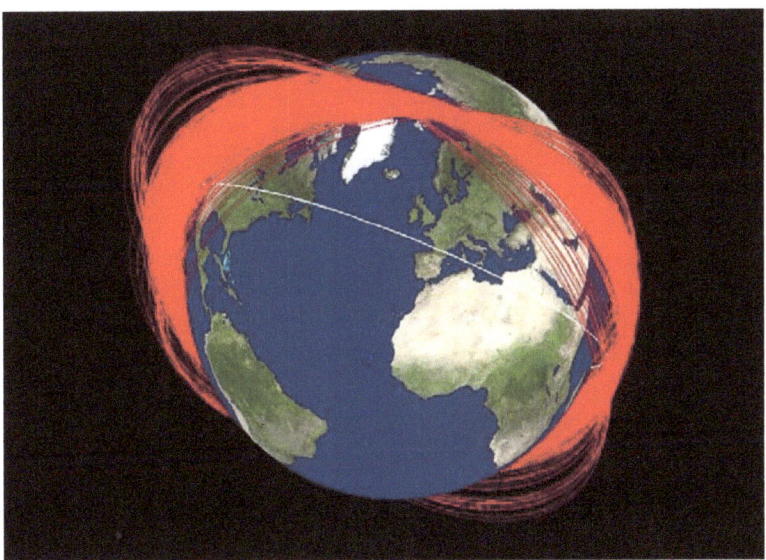

Figure 6:   Known orbit planes of Feng Yun-1C debris one month after its disintegration by a Chinese interceptor. Credits: NASA Orbital Debris Program Office [RD2].

Unfortunately not long after the Feng Yun-1C incident, another disaster took place. On 10 February 2009, the American Iridium 33 satellite, which was still operational, collided with the Russian non-operational Cosmos-2251 satellite at an altitude of 790 km. The two satellites hit each other with an angle of 90° and a relative velocity of just over 10 km/second (36,000 km/hour or 22,500 mph). They exploded into 2200 fragments bigger than 10 cm, increasing the total tracked debris population by another 16%, as can be seen in Figure 3. After the collision we see a decrease in total objects, since some of the debris fragments reentered in the atmosphere in the last five years. However by 2050, 10% of the original fragments will still be in orbit.

By now, the majority of Collision Avoidance Maneuvers by operational satellites in LEO are caused by the Feng Yun-1C and Iridium 33 / Cosmos-2251 fragments.

DEBRIS FALLING DOWN TO EARTH

Twice a day, a small space debris object reenters the atmosphere. Objects of moderate size, say 1 meter or higher, reenter every week. Normally this does not pose a risk to humans, as these objects enter the atmosphere with a speed of at least 29,000 km/hour (18,000 mph), and the dense atmosphere at lower altitudes start to slow down and heat up the debris, until finally the aero thermal heating burns up the debris before it even touches the ground.

However, some materials such as titanium or stainless steel have a high melting point and if large and compact satellites contain systems with these materials, such as fuel tanks, these objects do survive reentry and fall to the ground. Up till now, there are no reported cases of people being injured by falling space debris. Very large objects reenter the atmosphere only a few times a year and the Earth is covered for 75% by water and even on the land mass large parts are uninhabited, so chances of getting hit by space debris is small. In fact, the chance of being hit by lightning is bigger. Still...the chance is not zero and we see an increase in objects falling on Earth over the years. At some point in time, someone will get hurt.

The situations described above are known as 'uncontrolled' reentries. There is no control on where above Earth the object reenters, since the object is not functioning (such as navigation systems to guide the spacecraft towards the right position). A known example is the uncontrolled reentry of the Skylab space station in July 1979. Skylab is shown in Figure 7.

Figure 7:   Skylab in space. Credits: NASA.

The 74 tons station was so large that certain parts of it were bound to survive the reentry, and no one knew where the station would reenter. The station reentered above the Indian Ocean and Australia and several residents saw dozens of colored stripes through the sky. Many pieces crashed in Australia, such as the oxygen tank shown in Figure 8, but no one got hurt. Australia's Shire of Esperance however fined NASA a few hundred dollars for littering. A fine that was never paid until a radio station decided to raise a fund to pay the bill.

Figure 8:   Skylab Oxygen tank that crashed in Australia. Credits: Magnus Manske [RD3].

When the object that reenters carries dangerous substances, such as radioactive uranium or plutonium, the situation gets even worse. Several incidents have occurred in the past; the first in 1964 where the Transit 5BN-3 satellite reentered above the Indian Ocean after a rocket failure, releasing a kilogram of plutonium into the atmosphere. The last incident occurred in 1983 when Cosmos-1402 reentered over the South Atlantic, carrying radioactive uranium. Plutonium and uranium can be used as radio-thermal generators to generate power, also known as nuclear power. This makes satellites independent of solar power and requires no solar panels being pointed towards the Sun. Even Transit-4A carried nuclear power, but luckily only the upper stage exploded instead of the satellite itself.

The most widely known case of radioactive material crashing on Earth is when, in January 1978, the Cosmos-954 satellite equipped with a nuclear reactor, reentered the atmosphere above Canada in an uncontrolled way, and scattered pieces containing 30 kg of radioactive uranium over Canada along a swath of a 1000 kg. One can imagine that the Canadians were not very pleased with their country being sprayed by radioactive material. More on this topic will be discussed in the section on 'Legal matters of past accidents' in chapter 2.

However a reentry is not always uncontrolled. The Gemini and Apollo capsules returning to Earth were pinpointing to exact locations in the ocean, so that they could be retrieved by ships. Moreover, these capsules contained a Thermal Protection System (TPS) that prevented the capsule from burning up so that the capsule splashed down on the water intact with the crew alive.

Large objects like the Russian MIR space station and the European ATV (Automated Transfer Vehicle) supply vehicles to the ISS (see Figure 9), also reenter in a controlled way. Due to their mass, shape, and materials on board, some pieces are bound to survive the reentry and therefore they are targeted to reenter above an ocean to make sure no people on Earth get hurt.

Figure 9: Artist's impression of an ATV supply vehicle reentry. Credits: ESA.

COLLISION AVOIDANCE

If a dangerous object is coming your way, it is best to move aside a bit in order to avoid a collision. This principle applies when you ride a bike, but it also applies to satellites. In order to do this, we need to:

- know exactly the trajectories of the two objects that are about to cross each other's way
- have a propulsion system (well, one of the satellites needs it)
- be functional (the same satellite that has a propulsion system)
- have sufficient time (e.g. 12 hours or more)

Calculating the exact trajectory of all 17,000 objects for each second in the day is currently very difficult due to the computing power required, however there is a method in place that works most of the time: the SSN tracks all 17,000 objects and publishes 'two-line elements' for each object. Some owners and operators supply directly their two-line elements. These two-line elements are basically two lines of 69 codes, containing information on the position and velocity of the object in the form of orbital elements, and some data in order to propagate this position further in time. Two-line elements of objects can be found in [RD4]. From these data sets, we can quickly predict, for example within the next few days, close approaches using special trajectory propagators.

If there is a close approach, for example less than 300 m or if a collision could occur with a higher probability of 1 over 10,000, a collision warning can be issued and a more detailed analysis needs to be done. This involves obtaining more accurate data about the two objects, and then performing a more accurate propagation to predict if indeed the objects are approaching at close distance (say, less than 20 m for a satellite or 200 m for the ISS). If this is true, then a CAM needs to be performed.

The fact that this does not always work, is proven by the Iridium 33 / Cosmos-2251 collision. While from the two-line elements it was shown that a close approach was to occur at a distance of 117 m to 1.812 km, there were over ten other close approaches that day with even closer predicted distances (from the two-line elements sets). Therefore this close approach, which in fact turned out to be a collision, did not even show up in the top ten close approaches that day.

15

Figure 10: Close approach at the Moon! The LADEE satellite crossed within 9 km underneath the Lunar Reconnaissance Orbiter (LRO). LRO was able to capture LADEE in a picture where LADEE can be seen as a smeared line. Credits: NASA.

A CAM basically lowers (or increases) the altitude of one of the two objects by a few kilometer, until the objects have crossed, and then re-adjusts the orbit to its original state. This is a nuisance to operators as they have to spend time on planning, they may have to adjust the orientation of the satellite meaning that some sensors may not look into the right direction anymore and the satellite is not performing its normal function during this time.

The ISS performed its first CAM on 26 October 1999 to avoid a HAPS upper stage. While the HAPS had performed a fuel depletion maneuver (as described in the Explosions section before) which also lowered the orbit such that it would reenter within 25 years, it had a predicted close approach of less than 1 km. After a re-orientation of the ISS and a 5-second burn, this miss distance increased to 140 km. The ISS continued to perform CAMs once a year until the debris clouds of Feng Yun-1C and Iridium 33 / Cosmos-2251 lowered enough to cross the orbit of ISS, leading to two extra CAMs a year to avoid debris from that cloud, giving a total of three CAMs a year. Until early 2014, 17 CAMs have been performed by ISS.

Satellites located in orbits with more space debris than in other orbits will need to make more frequent CAMs. ESA's Sentinel-1A satellite had to make a CAM a few hours after launch! Sentinel-1A had not even reached its nominal pointing mode yet and got a confirmed collision warning with an old NASA satellite called ACRIMSAT (Active Cavity Radiometer Irradiance Monitor Satellite), to pass within 20 meters. Several vital subsystems on board had not been tested yet. During the night of 4 April 2014 a command was send to the satellite to execute the maneuver outside of reach of the ground stations. The maneuver was successful but certainly gave the operations team a night to remember. Collision warnings occur on a weekly basis in busy orbits, and it is not uncommon that satellites in busy orbits need to make a few CAMs each year.

COLLISION AVOIDANCE BETWEEN DEBRIS OBJECTS

In the previous section the collision avoidance method between an operational satellite and a space debris object was described, but what if two non-operational debris objects are on a collision course? None of the two will have the capability to maneuver aside, so can we prevent the collision by taking immediate action on Earth?

The first thing to analyze would be if it is a 'false alarm'. Trajectories of debris objects can only be determined to a certain accuracy, which means that we cannot truly determine if a collision is going to take place, we can only give a probability of collision. If the accuracy of the orbit prediction is low, such as 50 to 200 m, we may end up with many alarms. Today many alarms are triggered every day, while most of the time they are false alarms. Only if we can predict debris trajectories with accuracies of a few meters or maximum 10 m, we can truly predict, with high probability, if a collision is going to occur. This means that all satellite tracking mechanisms such as ground radars available today would need some serious upgrading. This is an expensive task in itself.

Once we have an accurate form of tracking space debris, and an alarm is triggered, how do we avoid the collision? The idea of 'shoot it out of the sky' clearly will not work as we have seen with the Feng Yun-1C experiment. This will actually ensure a collision and only create more debris.

A possible solution would be to give one of the debris objects a little 'nudge' so it deviates from its nominal trajectory. This could be done by launching a rocket towards one debris object, and let it softly impact the debris so it moves away. The rocket will need to have a very advanced navigation system to get exactly to the debris object within a short time (remember that a collision warning is only announced shortly before impact).

17

A kind of air cushion can be used to soften the impact with the debris. Once the debris is hit and moved away, the impacting rocket needs to move away as well, to ensure that it also does not collide with the other debris object.

A cheap rocket will need to be used, as it needs to be launched a number of times a year in order to avoid collisions. The navigation sensors and required software will be expensive, however many recurring hardware and software solutions will quickly lower the price. Still, multiplying a number of launches per year, by many decades of operations, as well as the radar upgrades on ground, will lead to a billion dollar investment just to avoid collisions between debris objects, and one can wonder if tax payers would be willing to invest in such a system, that in the end does nothing towards cleaning up space.

A system like this simply tries to prevent the situation from getting worse. It does not do anything to improve the environment. Collisions will still happen, as a system like this cannot have a 100% success ratio, so the debris population will still continue to grow, and the frequency of collisions will continue to increase. This, in turn, will ask for more frequent impact launches which in the end only drive the cost up.

SHIELDING

So far we have been discussing collisions of large objects, but satellites in space are frequently bombarded with tiny pieces of debris, say 1 mm or 1 cm thick. These pieces cannot be tracked from ground but with the help of recovered material from space (such as solar panels) or dedicated debris measurement satellites, plus debris environment models, we can estimate the total numbers. The result is that an estimated 750,000 objects larger than 1 cm orbit the Earth, and more than 166 million objects larger than 1 mm.

The Long Duration Exposure Facility (LDEF) was a large satellite that stayed in LEO for eight years before being retrieved by the Space Shuttle. Researchers found out that LDEF was impacted millions of times during its eight years in orbit. Some of the impact craters were visible to the naked eye; others were only discovered by using an electron microscope.

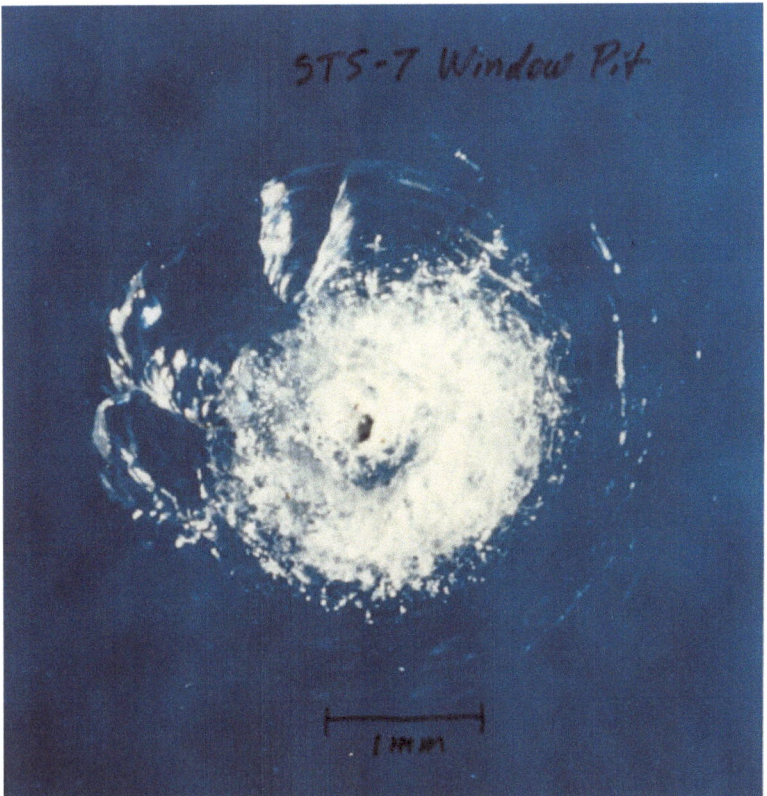

Figure 11: Debris impact crater on Space Shuttle window. Credits: NASA.

The Space Shuttle had windows and is therefore even more vulnerable to impacts of tiny space debris particles. Analysis of Space Shuttles returning to Earth has indicated that the windows are often hit. Over 70 Space Shuttle windows have had to be replaced. While the windows were not the weakest point, the replaced windows offered another source of data on impact fluxes, as shown in Figure 11. Another interesting point is that the chance of a window being hit depends on the orientation of the shuttle. In Figure 11 the velocity is aimed towards the right and we can see that for attitudes of the Shuttle with the window facing the direction of the velocity, the chance of window replacement is higher. We also see that if the tail of the shuttle faces space, the chances of window replacement is higher, though the effect is less strong than facing the velocity direction.

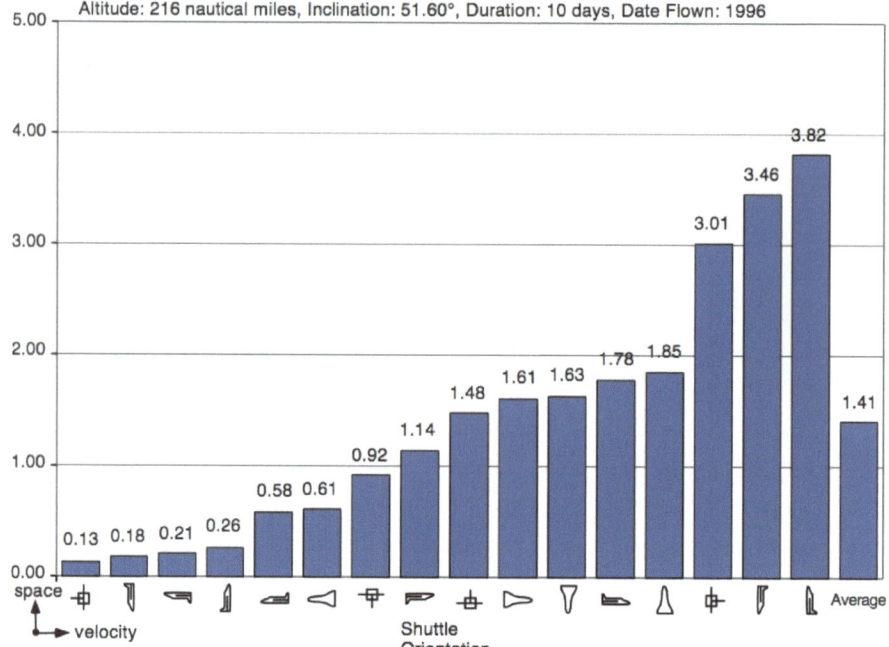

Figure 12: **Expected number of Shuttle window replacements as function of Shuttle attitude. Credits: NASA [RD5].**

This attitude dependence also teaches us that we can protect sensitive equipment by moving them away from the most probable impact direction, during the design of a satellite.

Several forms of shields are available, such as the Whipple shield, which is basically two layers of aluminum: one bumper and one back plate. The bumper breaks up and partially vaporizes the incoming particle, spreading out the impulse over a larger area on the back plate, see Figure 13. This design was proposed in 1947 by Fred Whipple.

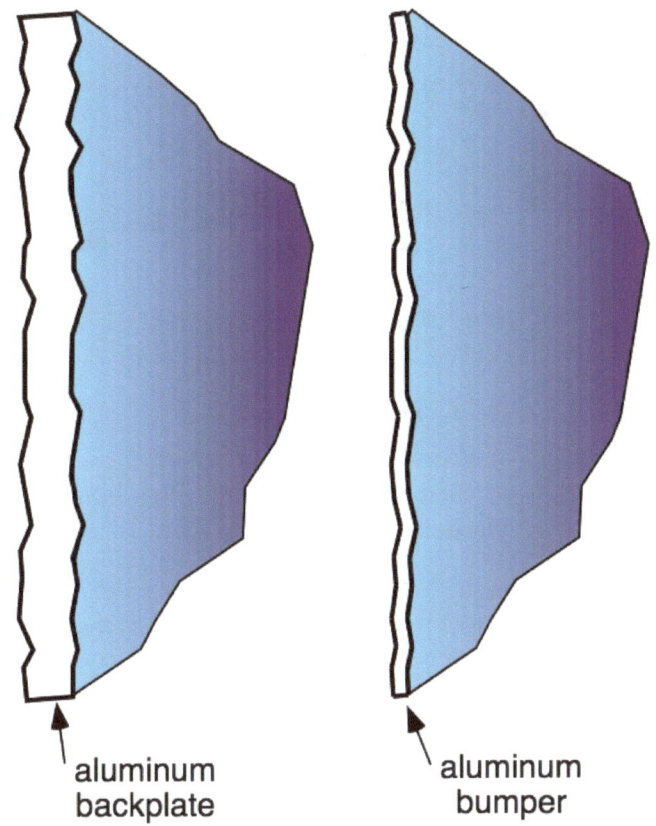

aluminum
backplate

aluminum
bumper

Figure 13: Whipple Bumper shield. Credits: NASA [RD5].

Several variants exist of the Whipple Bumper, such as the Multi-Shock shield which replaces the bumper with several layers of ceramic fiber, or the Stuffed Whipple Shield which combines the two and contains a blanket comprising of multiple layers of aluminum mesh and ceramic fabric, similar to a bullet-proof vest. The latter is used on the ISS. Figure 14 shows the Stuffed Whipple Shields on the European Columbus module attached to the ISS. On the ISS, there is typically 10 cm between the back plate and the bumper.

S122E008263

Figure 14: Whipple Shields on ESA's Columbus module of the ISS (cylinder shape). Credits: ESA.

These shields do not only protect against tiny space debris particles: they also protect against impacts of micrometeorites. However, the protection only applies to small particles, up to an impact of a 1 cm object. A larger object will penetrate the shield and the body of the satellite. This may not necessarily lead to a complete disintegration of the satellite, but for example, being hit by an object of 10 cm or larger with an impact velocity of 36,000 km/hour certainly will, as the released energy is similar to an exploding hand grenade. There is certainly no protection against a satellite of hundreds of kilograms hitting you at that speed.

An alternative to outside shielding is placing vulnerable equipment in specific locations inside the satellite such that other components are shielding these vulnerable components. Fuel tanks are good shielding for components for example. Another example is the satellite's central cylinder (if present). Often fuel tanks are placed inside this cylinder, but other sensitive equipment could be placed inside as well. Furthermore, a vulnerability assessment could be done that analyses which direction is dominant during the mission in terms of small debris impacts. By placing sensitive equipment on the opposite side, they can be shielded from incoming debris by other equipment such as fuel tanks.

CHAPTER 2:   OWNERSHIP OF SPACE DEBRIS

We have seen that the amount of space debris in Earth orbit is increasing rapidly, and that operators on a regular basis now have to dodge space debris. We also know that we can apply shields to protect us from the millions of tiny particles floating around; however the large objects pose a problem. With the knowledge that only 600 of the 170,000 tracked objects in space larger than 10 cm are functional satellites, we can think of removing the majority of the objects that are space debris. However are we allowed to do this? And what do we do in the future to prevent more debris creation?

LAWS GOVERNING SPACE DEBRIS

The space debris problem is a problem that belongs to us all. It is an intergovernmental issue and therefore it makes sense to look at an intergovernmental organization that looks at protecting the environment: the United Nations (U.N.), including its Committee on the Peaceful Uses of Outer Space (COPUOS).

Figure 15:   The first woman and man in space, Valentina Tereshkova and Yuri Gagarin, visit the U.N. in 1963. Credits: U.N.

In 1967 the U.N. opened for signature the "Treaty on Principles Governing the Activities in the Exploration and Use of Outer Space", also known as the **Outer Space Treaty**. It entered into force the same year, and by now 102 countries are parties to the treaty, including the US, the former Soviet Union, the United Kingdom, China and India. The Outer Space Treaty defines the principle of no ownership in space, such as natural resources and celestial bodies. Furthermore it states that a state to the party that launches an object in space, or a 'launching state', has jurisdiction over the object.

The term 'launching state' is defined in the U.N. 1972 "Convention on International Liability for Damage Caused by Space Objects", also known as the **Liability Convention**. While the state that launches a rocket is considered a launching state, an international organization procuring a launch, like ESA, could also be launching state (ESA, in fact, is a launching state). This means that if an ESA satellite is launched by a French Ariane 5 rocket, the Ariane 5 upper stage in space belongs to France, since the Ariane 5 was launched from French Guiana, while the satellite in orbit belongs to ESA. The Liability Convention states that space objects legally belong to the nation state that launched them, and is internationally liable for damage to another state party to the treaty. In the Liability Convention space objects are formally defined as: "including component parts of spacecraft, their launch vehicles, and component parts of their launch vehicles", meaning that space debris clearly fall under the space objects to which the Liability Convention applies.

Finally, a convention worth noting is the 1976 "Convention on Registration of Objects Launched into Outer Space", also known as the **Registration Convention**. It requires all launching states to register newly launched space objects with the U.N. In reality, this is not always done and there is no fine for not doing so, so one could wonder if a launching state could simply not register an object and therefore not be liable for it. Of course for large objects like Cosmos satellites or Ariane upper stages, the ownership is clearly defined. Moreover Article VI provides a procedure for determining the identification of an object if it is unregistered and caused damage to others.

To summarize, several aspects are applicable to the removal of space debris from orbit:

- **No nation can claim that debris exists exclusively in its territory**
- **A launching state that puts an object in space has and keeps jurisdiction over that object, even when it turns into space debris**
- **Debris can therefore only be removed with the consent of the launching space that put it into orbit**
- **The launching state is liable for damage caused on Earth from either the launch or reentry of the object**
- **The launching state is liable for any in-orbit collision if fault can be established.**

So what happens if a launching state gets approval to remove a space debris object belonging to another state, and then something goes wrong and the debris reenters in a wrong place, causing damage in orbit or on ground? And perhaps even injury or casualties on ground? Some serious rules will need to be established since the original launching state of the debris probably prefers not to be liable for mistakes made by the owner of the debris remover. Of course if a launching state removes its own debris, these issues are mostly avoided since there will not be a transfer of liability. It is very likely that the first debris removals will concern launching states removing their own debris. Figure 16 gives an overview of the distribution of debris in terms of ownership.

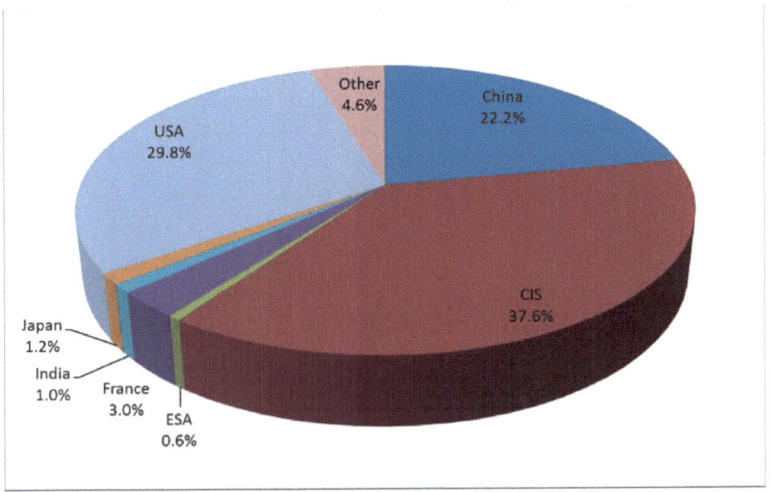

Figure 16: Debris ownership as recorded in January 2015. Data source: [RD9].

One aspect is missing in the above bullet points though, which may be one of the most important aspects of all:

There is no current international legal duty (yet) to remove space debris from orbit.

The Outer Space Treaty does mention "In order to promote international co-operation in the peaceful exploration and use of outer space, States Parties to the Treaty conducting activities in outer space, including the Moon and other celestial bodies, agree to inform the Secretary-General of the U.N. as well as the public and the international scientific community, to the greatest extent feasible and practicable, of the nature, conduct, locations and results of such activities" which is more of a guideline. The next section goes more into detail on existing guidelines.

ADOPTED SPACE DEBRIS MITIGATION RULES

Lubos Perek was born in Czechoslovakia in the year 1919. After graduating from university he started working on his PHD degree on Astronomy at Charles University in Prague. Following his university career he worked himself up and became chief of the Outer Space affairs Division at the U.N.. In 1977 he presented a paper called "Physics, Uses and Regulation of the Geostationary Orbit, ex facto sequitur lex". The presentation describes aspects of the environment in the Geostationary orbit (GEO), as well as how this can create problems. More importantly, this paper is among the first to discuss the issue of orbital debris in GEO.

Two years later he published and presented another paper called "Outer Space Activities versus Outer Space". In this paper Lubos Perek is the first to recommend space debris mitigation measures. These included to reduce the amount of debris produced during launch and operations and to deorbit inactive satellites. Furthermore he proposed to move GEO satellites into a disposal orbit at a bit higher altitude than the Geostationary orbit. His recommendations are still applicable today.

In February 2007, 28 years after Lubos Perek's paper on space debris mitigation measures, the U.N COPUOS adopted the following space debris mitigation guidelines:

1. Limit debris released during normal operations

2. Minimize the potential for break-ups during the operational phases

3. Limit the probability of accidental collision in orbit

4. Avoid intentional destruction and other harmful activities

5. Minimize potential for post-mission break-ups resulting from stored energy

6. Limit the long-term presence of spacecraft and launch vehicle

7. Limit the long-term interference of spacecraft and launch vehicle orbital stage with geosynchronous (GEO) region after the end of their mission

These guidelines were adopted one month after the infamous Chinese anti-satellite test of Feng Yun-1C, but represent a long effort by the IADC in which for example NASA and ESA have representatives. Agencies like these were using space debris mitigation measures for years already. ESA, for example, had a space debris mitigation handbook since the year 2000 and has been organizing workshops on space debris since 1987. NASA had similar guidelines and organized the first international major conference on space debris in 1982 at the Jet Propulsion Laboratory. However the guidelines stayed nothing more than guidelines for a long time. Only in 2008 were the space debris mitigation guidelines adopted as a policy by ESA, for example. The International Organization for Standardization (ISO) adopted in 2011 ISO 24113; an ISO standard that makes recommendations on space debris. This ISO standard has been adopted as policy for ESA in 2014. Agency policies on space debris mitigation are therefore a recent thing; before 2010 the debris mitigation rules were nothing but guidelines, endorsed by the U.N. and national/international space agencies.

France, being a launching state, has adopted the 'French space operations act' (FSOA) which is a law at national level, with very similar rules as stated in the ISO 24113, as well as stating the maximum casualty risk.

From the collection of standards, laws and regulations, we can gather two rules which are of most importance for satellites in LEO:

1. Satellites shall remove themselves from protected zones in space within 25 years after the end of their missions. The probability of a successful disposal shall be at least 90%.

2. Satellites that reenter into the atmosphere shall not pose a risk of fatality on ground of more than 1 in 10,000.

To implement the first rule we must first know what the protected zones in space are. The next chapter will explain this. A second aspect of this rule is that it does not state where the debris shall go to; it only states to remove debris from the protected zone. It does not state the destination, or the method.

An accurate prediction of the reentry time is almost impossible. The Earth's atmosphere is volatile and expands to higher altitudes (and thickens at lower altitudes) when the Sun is more active, which happens every now and then. When this happens satellites experience more drag and will reenter quicker. When the Sun is less active, the drag force will be lower and then satellites take longer to reenter. If satellite designers assume a thick atmosphere, the satellite may not reenter in 25 years if the Sun becomes less active. And if they assume a thin atmosphere, more fuel is needed for the deorbit maneuver at the end of their operational life because they need to lower the altitude more than when assuming a thick atmosphere. A compromise may be required that assumes an average solar activity.

The second important rule is that satellites reentering shall not have a fatality risk higher than 1 in 10,000 on ground. Several questions come to mind when a rule like this appears: when will the satellite reenter? What will be the world's population at that time? How do we calculate that probability? And what do we do if the chance of a casualty on ground is higher than 1 in 10,000?

Satellites in high orbits, such as 800 km or higher, may take hundreds of years to reenter, unless the satellite has performed a deorbit maneuver at its end of life. Typically we could expect reentry time between 25 and hundreds of years. It will be difficult to predict the world's population in hundreds of years. Population growth equations do exist, such as the world population prospects published by the U.N., as shown in Figure 17 below, but may not hold for hundreds of years; especially since different models lead to different results. Currently objects of medium size are entering our atmosphere on a weekly basis. With the world being covered for 75% by water, the chance of being hit by space debris could not possibly be higher than 25%, or 1 in 4. However even large portions of land masses are uninhabited, making the probability even less. The ground track of the satellite has a large influence on the probability of casualty. A ground track is basically the track of the satellite as it orbits space, but projected on to the surface of the Earth. Standing on the ground track of a satellite would mean that this satellite will pass right over your head.

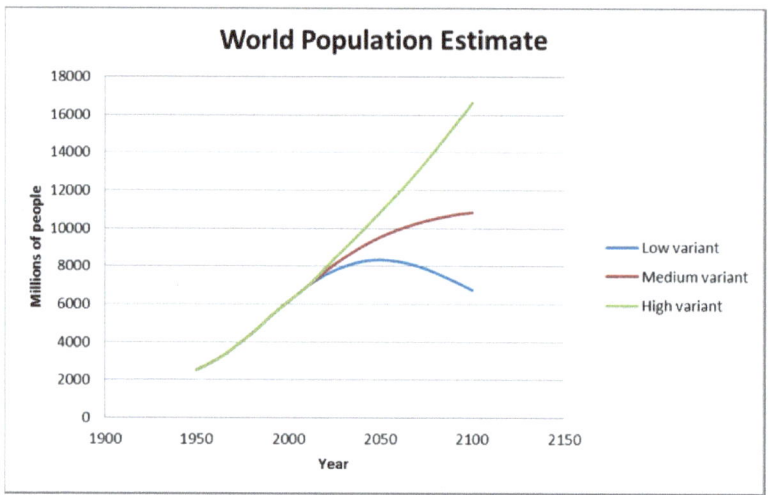

Figure 17: World population estimate by the U.N..Data Source: [RD10].

Figure 18 shows a ground track of the ISS. You can see that the ISS often spends time over inhabited areas. The arctic and Antarctic areas are normally uninhabited however the ISS never crosses the 52 degrees latitude circle. Instead it spends a lot of time in the 30 to 50 degrees latitude band, both North and South, which is typically well inhabited.

Figure 18:   Ground track of the ISS [RD8].

If we now look at the ground track of ESA's Sentinel-1A satellite (Figure 19), we see that it spends much time over the polar areas. As a result, if this satellite would reenter, the chance of it coming down over an uninhabited area is much higher, and therefore the casualty risk lower.

Figure 19:   Ground track of ESA's Sentinel-1A satellite [RD8].

If we know the ground track, we can simulate the possible projected ground area of debris that survives the reentry. If we divide the area of the population by the area of impacting debris, we have the probability of casualty on ground, assuming for example that one human occupies a square meter. However how do we know what survives the reentry and what not? Since it is heat transfer that burns up satellites entering our atmosphere, we can expect materials with high melting points to have a chance of survival. Examples are titanium, carbon ceramic or stainless steel. We have seen in the past that large fuel tanks are prone to survival. Also large synthetic aperture radars could be a problem. Heat shields (with a TPS) would be designed to survive such heat.

Other aspects that influence the reentry survival are how compact the satellite is and how large it is. Compact and heavy satellites have a higher chance of debris survival than large but light satellites. If a satellite has many appendices, or panels sticking out (and is therefore not very compact), these parts are likely to break off during reentry and burn up. The attitude during reentry matters here, and we could think of future satellites being designed to enhance the chance of break-up during reentry, as well as the choice of materials. This is called 'design for demise'.

A larger satellite leads to an increase in the area of its surviving debris falling on ground. Simulation and past space debris objects found on ground have shown that a typical rule of thumb is that satellites heavier than 1000 kg will have a higher risk of casualty on ground than 1 in 10,000. Small satellites (200 kg or smaller) will have a lower risk than 1:10,000. For satellites in between 200 and 1000 kg, simulations will need to be performed, taking into account the orbit, shape, materials used, configuration of the satellite, and population models, in order to calculate the risk of casualties on ground.

And in case the risk of casualty is higher than 1:10,000, how can we lower this risk for a satellite that is already in orbit? The only way to do this is by ensuring that the satellite reenters over an uninhabited area, such as an ocean. If the satellite is in design phase, a propulsion and guidance system needs to be added that allows the satellite to perform a precision maneuver such that it reenters over an ocean.

If a large satellite is already in orbit but cannot perform such a maneuver itself, because it lacks a propulsion system or because it is not functioning anymore, the only solution is to have it removed by using a space garbage truck...

LEGAL MATTERS OF PAST ACCIDENTS

We can consider four big incidents in history related to space debris, as described in the previous chapter:

1. The Feng Yun-1C destruction by a missile

2. The collision between Iridium 33 and Cosmos-2251

3. The reentry of Skylab on Australia

4. The reentry of Cosmos-954 with nuclear reactor, over Canada

The Feng Yun-1C incident itself cannot lead to a settlement in court as it concerns two objects of the same country. It is therefore not an international incident as such, implying no international liability claim under the U.N. Liability Convention. Nevertheless in the future, an operational satellite may be hit by debris from this event, as the debris cloud from this event is giving collision warnings on a weekly basis.

The collision between Iridium 33 and Cosmos-2251 involved satellites launched by two different states. In terms of launching states, Iridium had several launching states: the USA (which procured the satellite via its company Iridium), the Russian Federation (which leased the Baikonur launch site) and

Kazakhstan (which owned the territory from which the Russian Federation leased the launch site). However since Iridium 33 was under jurisdiction and control of the USA, the USA holds liability. For Cosmos-2251, the Russian Federation was the launching site. Cosmos-2251 was launched from the Russian territory Plesetsk. Neither USA nor Russian Federation registered their satellites with the U.N. as required by the Registration Convention.

The Liability Convention states that a launching state is liable only if the damage is due to its fault, or the fault of persons for whom it is responsible. Unfortunately, we lack a clear definition of the term 'fault'. Did the Americans or the Russians do anything faulty at the time of collision? The USA was simply operating their satellite, and Cosmos-2251 was out of order. Both USA and Russian Federation were therefore doing nothing but business as usual, which does not look wrong. However a fault can be caused by negligence, and this is often the meaning used in international law. Was the USA negligent by not monitoring collision risk properly? There is no collision risk monitoring in Iridium. the SSN does monitor collision risk but with a focus on military satellites. It also publishes two-line elements of all cataloged objects, but the orbital elements inside two-line elements are not accurate enough to accurately predict a collision. In fact, calculations based on two-line elements showed that Iridium 33 and Cosmos-2251 did not even show up in the top ten of highest probability collisions. Still, the fact that Iridium did have the means to perform a CAM, and the USA does have the means to accurately follow Iridium's orbit, negligence is shown.

Likewise, negligence can be shown in the Russian Federate, simply for leaving a massive satellite out there. This was not an active mistake since no laws claim that they should have removed their satellite from orbit. However it can be considered negligent. Even if both parties did not register their satellites with the U.N. under the Registration Convention, which does not relieve them of possible liability. However compensation can only be claimed if the other party proves fault on the other part. Also, the severity and nature of the damage it has suffered needs to be proven. The USA could have prevented the collision if it would have monitored Iridium 33 and executed a CAM, but they were the ones losing an expensive and operation satellite part of a larger constellation. The Russian Federation could have done nothing to prevent the collision, but suffered no real damage as their Cosmos satellite was already not functioning anymore. Finally, no compensation was claimed from either side.

The Skylab reentry had a casualty risk higher than 1:10,000. In fact the calculated casualty risk was 1:152. Skylab reentered years earlier than expected due to high solar activity which thickened the atmosphere. While NASA aimed at a reentry South of South Africa, it actually came down over South-West Australia with large fireballs in the sky and sonic booms waking up people early in the morning. 500 pieces of debris weighing 20,000 kg in total fell down over a footprint of almost 1000 km long and 200 km wide. While the reentry was clearly dangerous and created awareness that space debris could pose hazards to people on ground, no international claim was filed by Australia to the USA, via the U.N. Only NASA got fined for littering, as mentioned in the previous chapter.

The reentry of Cosmos-954 over Canada resulted in the first application of the Liability Convention. The satellite carried a nuclear reactor based on uranium-235. The nominal mission should have ejected the reactor and left it in an orbit from which it would not reenter within the next 300 to 1000 years (meaning the intention was simply to shift the problem to next generations, as the uranium would still be radioactive at that time). However this action failed and when Cosmos-954 reentered, the nuclear reactor reentered with it, spreading radioactive debris over a region of over 600 km long in Northern Canada. A $14 million recovery operation was performed (called 'Operation Morning Light') and since there was a clear fault on the Russian side, and clear damage on the Canadian side, Canada claimed a $6 million compensation from the Soviet Union under the Liability Convention, of which $3 million was eventually paid.

CHAPTER 3: HOW TO CLEAN UP SPACE

We have now learned the rules when it comes to debris removal. If we are to launch a space garbage truck, the debris mitigation rules also apply to this spacecraft. We will have to make sure that (1) it is removed within 25 years from the protected zones in space, and (2) if we are to remove an object by reentry, and/or we reenter ourselves, the probability of casualty on ground shall not be higher than 1:10,000. We can translate these points into mission requirements for the space garbage truck mission. For example:

> MISREQ1 = "Remove a large space debris from the LEO protected zone"
> MISREQ2 = "All elements in the mission shall be removed from the LEO protected zone within 25 years
> MISREQ3 = "The mission and all its element shall not pose a probability of casualty on ground higher than 1:10,000"

These requirements have a strong impact on the reliability of the components on board, as well as on the accuracy of the guidance system. Finally, we learned that we will need approval by the launching state that is liable for the debris, if we want to remove a certain debris object. In fact, this is not too far off with respect to our garbage on ground: it is generally not appreciated if we take garbage from our neighbor's garbage container, and move it somewhere else. Even if we would actually clean it up, you would still want to ask your neighbor first...

Let's have a look at a typical sequence of a garbage truck daily routine for garbage removal. Figure 20 gives a functional diagram of a garbage truck on Earth.

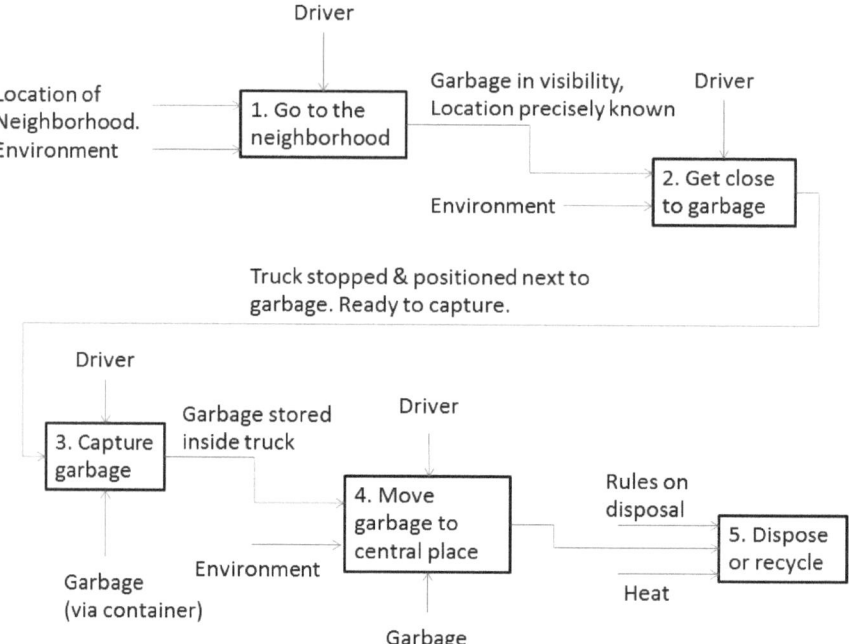

Figure 20: Functional diagram of a typical Earth garbage truck removal, showing functions and interfaces

The functions are described as blocks in 'Integrated Definition' (IDEF) format, meaning that the blocks have inputs coming in on the left side, create outputs coming out of the right side. It shows that a removal consists of:

1. Getting to the neighborhood of the garbage collection

2. Getting precisely to the garbage

3. Capture the garbage

4. Move the garbage to a central place

5. Dispose or recycle garbage

Commands that influence the functions are coming in from the top of the block, and external interfaces are connected to the bottom of the block.

In this chapter we will take a look at how we can remove debris: how to locate it, and how to remove it. We will focus therefore on functions 1, 4 and 5. Functions 2 and 3 will be discussed in the next chapter.

WHERE IS SPACE DEBRIS?

Let's start with function 1 in Figure 20: getting to the neighborhood of the garbage collection. On Earth, the truck (see Figure 21) is driven by a driver, who commands the vehicle using the steering wheel, accelerator and hopefully the brake as well. This is seen as 'the command' which enters from the top of the block. The driver steers the truck to the neighborhood which has a dedicated place on a map that the driver is aware of. As inputs we have the environment of the Earth: the roads that the driver needs to follow, the speed limits to which he/she needs to keep. The output is that the truck will be close to the garbage to be collected, i.e. 'in the neighborhood'. The driver will drive at reduced speed because normally the speed limit in living areas is lower than the speed limits of main roads leading to the living areas.

Figure 21: Typical garbage truck on Earth leaving to collect garbage. It features a robot arm on the side to catch the garbage container. Credits: Angelo T [RD6].

In space, the function is exactly the same: getting to the neighborhood of the space debris. There are two fundamental differences: first the environment, which is now space. Space has a strong impact on our space garbage truck, in particular in terms of the orbits it can follow and the temperature differences between sunlight and shadow. Secondly, there is no driver on board. Though there certainly could be, it would make the cost and complexity of the garbage mission very high, as putting humans on board requires a higher level of testing, reliability and redundancy. Also, several systems would need to be inserted to keep humans alive, such as oxygen supply, room to sit, controls, escape systems, and windows to help with the navigation in space when being close to the debris. If we focus on a robotic mission (meaning no humans on board), the spacecraft will still need to communicate with humans on ground. This is done via 'TC' which means telecommands, and 'TM' which means telemetry. So the spacecraft could receive a telecommand from ground to start the engines, and send back telemetry containing the confirmation that the engines have been started. Other telemetry is 'house-keeping data' which is information about health or status of the satellite: temperature of all the equipment, information about the memory, orbit data, are all examples of house-keeping telemetry.

If we have as input the location of the debris, we need to know where debris around the Earth is. Figure 22 shows a distribution of the density of space debris as function of altitude, within the LEO protected zone range (up to 2000 km). The spatial density is the amount of satellites per cubic kilometer. The upper curve is the most recent density, and there is a clear peak in the 600 to 1000 km altitude band. There is another smaller peak in the 1400 to 1600 km band. The ISS is typically orbiting around 400 km altitude and the spatial density is about 16 times less than the highest density.

The peak is close to 800 km, typically 760 to 780 km. The 800 km region is a region often used by Earth observation and military satellites. At this altitude there is little disturbance by the drag force as the atmosphere is extremely thin at that altitude. At the same time, the altitude is still low enough for sensors on board satellites (cameras, radars) to give high precision data. As a result, around this altitude there are many satellites which are now defunct or inactive, as well as many upper stages.

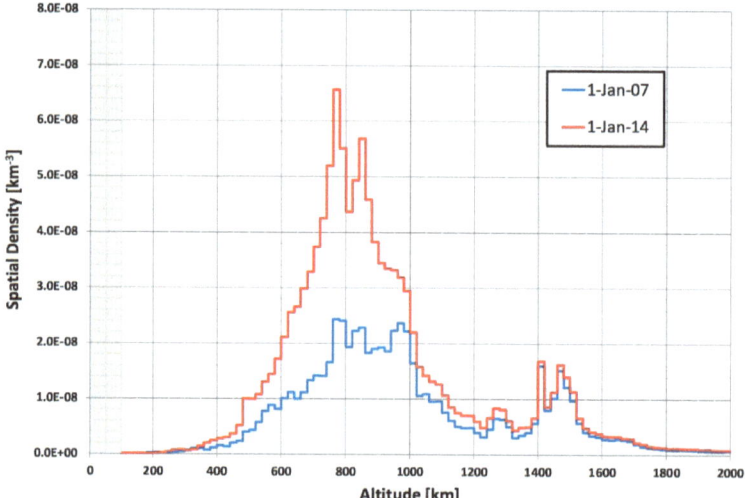

Figure 22:   Spatial density distribution of space debris at 1 January 2007 (blue lower curve) and 1 January 2014 (red upper curve). Credits: NASA [RD7].

We also see that the amount of debris in this area is over twice as much in 2014 as in 2007. The Feng Yun-1C and Iridium 33 / Cosmos-2251 incidents are large contributors to this. Figure 22 only talks about altitude and this does not give us a three-dimensional overview of the debris problem, so we do not know if we are talking about satellite that span lower latitude ranges (like the ISS as shown in Figure 18), or satellites that also cover polar regions (like Sentinel-1A as shown in Figure 19).

In order to do this, we introduce the orbit parameter called 'inclination' which stands for the angle between the equator and the plane in which the satellite orbits; see Figure 23. When the inclination is zero degrees, the satellite moves within the equator (it only reaches zero degrees of latitude). When the inclination is 90 degrees, the satellite moves in a 'polar orbit', since it moves from South pole to North pole and then back to South pole. If satellites need to cover the entire Earth, including the poles, they are required to be in polar orbits. The inclination can even be higher than 90 degrees and in fact many satellites in LEO have inclinations of 98 to 100 degrees. These are called 'Sun-synchronous orbits' (SSO) and have as main feature that the Sun has an almost fixed position with respect to the orbit plane. This has two big advantages: first of all the satellite has similar light conditions when looking down to Earth. A satellite could cross the equator (moving from South to North) every time at 10:00 in the morning for example, which allows good comparison of multiple observations since they were measured at the same time in the day. Another advantage is that in space, the position of the Sun does not change with respect to the orbit plane so it is easier to point solar panels to it. These two advantages make the SSO very popular, and much debris can be found in these orbits, for example the Sentinel-1A satellite of which the ground track was shown in Figure 19. An inclination of 180 degrees brings the satellite back into the equator plane, but now rotating around the Earth in the opposite direction to a satellite with an inclination of zero degrees.

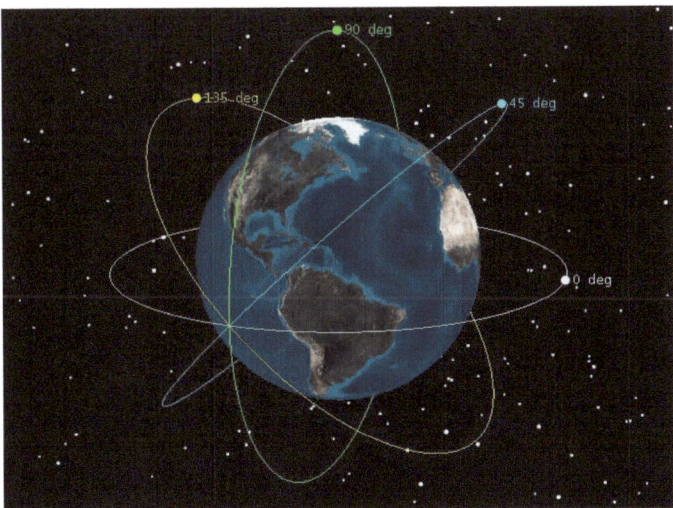

Figure 23:   Definition of the orbit parameter inclination [RD8].

As it turns out, most of the debris is situated around the polar orbits. Figure 24 shows an overview of the debris in terms of inclination. We can distinguish three locations: a band around the 70 degree inclination, a peak at 82 degrees and a peak at SSO inclinations (98-100 degrees). The around-70-degrees-inclination area is typically filled with large Russian upper stages, such as Zenith upper stages. In the SSO region we still find many operational satellites today.

So when we start cleaning up, we know now that the 'neighborhood' to go to is the 600 to 800 km altitude band and high inclinations (+- 70 degrees, 82 degrees and SSO).

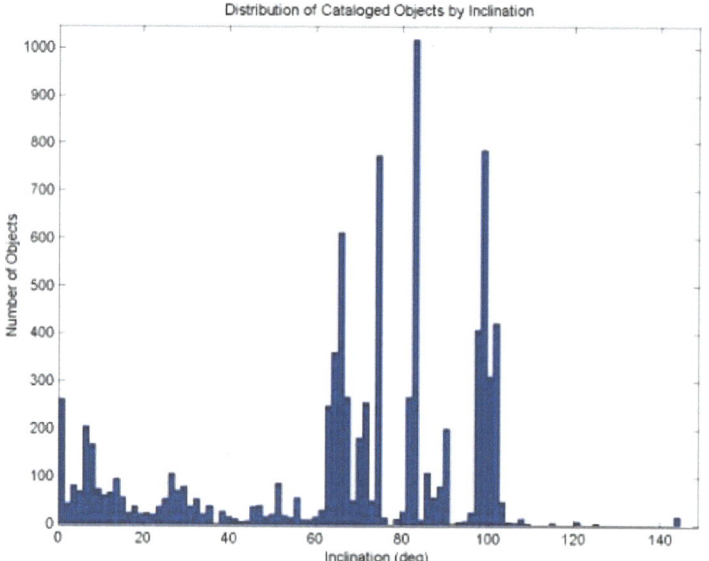

Figure 24: Distribution of space debris as function of orbit inclination. Credits: Zach Wilson [RD11].

Obviously, LEO is not the only place in space where space debris is located. Other orbits, such as GEO have debris too. However the majority of debris is located in LEO, and while the GEO orbit is regulated by ITU (International Telecommunications Union) which states that satellites occupying positions in GEO must be removed after their end of life, regulations for LEO have only come into existence recently as we saw in the previous chapter. For this reason we will focus on removing debris from LEO in this book.

THE PROTECTED ZONES IN SPACE

In the previous chapter, section 'Adopted Space Debris Mitigation Rules', two debris mitigation rules were defined. To implement the first debris mitigation rule ("Satellites shall remove themselves from protected zones in space within 25 years after the end of their missions."), or in order to understand where to dispose of the debris, we must first know what the protected zones in space are. Figure 25 shows these zones. Basically everything below 2000 km altitude is protected (known as the LEO zone), as well as a band spanning from 200 km below to 200 km above the GEO ring of 36,000 km altitude. Note that the band spans vertically in a plane from 15 degree below the equator plane of the Earth up to 15 degrees above the equator plane, while the protected LEO zone is basically a sphere around the Earth up to 2000 km. So the LEO zone covers all inclinations, while the GEO zone covers inclinations up to 15 degrees.

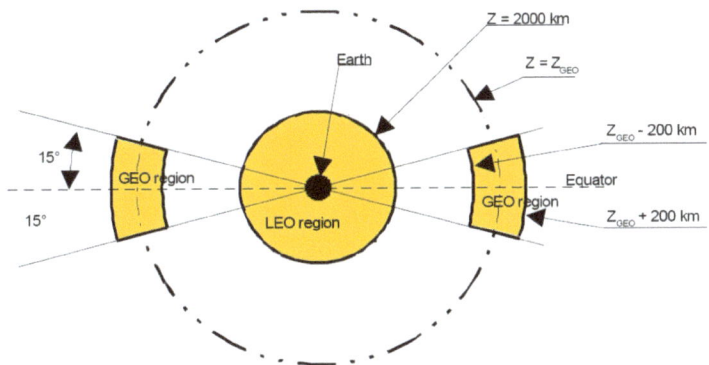

Figure 25:   Definition of the protected zones in space. Credits: ESA.

A second aspect of this rule is that the requirement is only to remove the satellite from the protected zone; not necessarily to reenter it. A LEO satellite could be moved up to above 2000 km for example. The same applies to GEO satellites. However without letting the satellite burn up in the atmosphere, we are not exactly cleaning up space by moving satellites to another unprotected zone in space. Also, it takes more fuel to move a satellite that orbits at an altitude of 800 km up to 2000 km, than it takes to move it down into the atmosphere. For GEO satellites, it would take an enormous amount of fuel to reenter them into the atmosphere from such an altitude (36,000 km), so normally GEO satellites are moved to at least 300 km above the GEO altitude. While there are no rules for unprotected zones, sometimes a best effort approach is applied to ensure that the satellite does not enter a protected zone within at least 100 years after the operational mission. ESA's GAIA mission is orbiting a point 1.5 million km's away from the Earth. A small maneuver will be executed at the end of the mission to make the satellite drift away from the Earth and escape the Earth's sphere of influence.

Finally, note that the first debris mitigation rule only states that debris is to be removed from the protected zone within 25 years. Most satellites that are orbiting at an altitude of 600 km or lower, will automatically comply with this rule. At this altitude the drag from the atmosphere is still present, and within 25 years the orbit will have lowered low enough for the satellite to enter the atmosphere automatically. It also means that satellites that orbit at an altitude of 800 km, could simply lower their altitude to 600 km to reenter in the atmosphere within 25 years for example (there are in fact other orbits that costs less fuel to achieve and will also reenter within 25 years).

REMOVING DEBRIS OUT OF THE PROTECTED ZONES

Now let's examine how to remove the debris from the protected zones in space. There are two ways to be removed: actively and passively. With Active Debris Removal (ADR), our space garbage truck would grab the debris and remove it. With passive removal, the debris would remove itself from the protected zone. However what happens if the debris is so large that the casualty risk is higher than 1 in 10,000? Even if the debris is removed from the protected zone within 25 years, we would have a violation of the casualty risk rule.

Historical data has shown that satellites up to 500 kg do not pose a large casualty risk: normally all material burns up in the atmosphere, and even if small pieces survive the casualty risk is below 1 in 10,000. For these objects, passive removal techniques will suffice. A device can be put on to the satellite that ensures that the satellite will reenter within 25 years after the nominal operations of the satellite. Then, both 25-year rule and casualty risk rule will be complied with.

For satellites of 1000 kg and higher, there is a high probability of violating the 1 in 10,000 casualty risk. For these objects, we must use ADR to ensure that the satellite reenters over an uninhabited area. This is done, for example, with ESA's ATV vehicles. After undocking from the ISS, a large propulsion system pinpoints the ATV down to reenter in the South Pacific. In case there are pieces surviving the reentry, these pieces will fall into the ocean and not on cities for example. Unfortunately though, as we have seen in the previous chapters, there are many pieces of debris in orbit, and many of them are heavier than 1000 kg. When those pieces enter the atmosphere, there is a higher chance than 1 in 10,000 of casualty on ground, each time such a reentry occurs. For these cases, a space garbage vehicle can be used to remove them from orbit.

When satellites are in the grey area, such as having a mass between 500 and 1000 kg, many simulations need to be done to see what the casualty risk on ground is when the satellite enters the atmosphere. This will depend on the shape, the size, the type of materials on board, and the orbit. It may be that a small satellite of 500 kg can still pose a large casualty risk simply because it has an orbit with a ground track over densely populated areas.

The following sections show two possible methods to remove small debris from orbit: drag augmentation devices and propulsive devices, as well as the method for large debris.

SATELLITE LIFE TIMES IN LOW EARTH ORBIT

Why do satellites in LEO lower in altitude over time? The simple answer is drag. A more extensive answer is: due to the drag forces that acts on the satellite, which depends on the density of the air, the shape of the satellite, the size of the satellite and the mass of the satellite.

The density of the air is a function of altitude and time. The altitude function is easy: the higher the altitude, the lower the density. See Figure 26 below for an example. The graph shows the density in logarithmic scale and does not take into account the temperature of the air, humidity and other things that change over time.

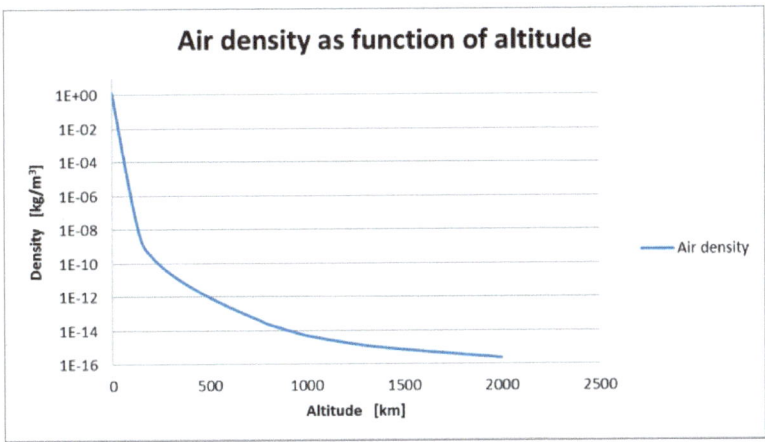

Figure 26: Average air density as function of altitude. In logarithmic scale.

At an altitude of 2000 km, the density is practically zero. And at the Geostationary orbit, there is simply no drag. However at 800 km, there is still enough 'air' to place a drag force on to the satellite that lowers the altitude by a few km per year. And at 600 km altitude, the atmosphere is considered thick enough to ensure that the satellite reenters within 25 years.

The shape is rather similar for every object in space: a cube for a satellite and a cylinder shape for a rocket upper stage. A cube is an optimal shape for putting equipment inside satellites, and store solar panels against the side walls. Unless satellites are designed to be aerodynamic (such as ESA's GOCE satellite which was orbiting the Earth at low altitudes), a cube is the most common shape of satellite.

Figure 27: ESA's GOCE satellite. Credits: ESA-AOES-Medialab.

The size and mass of the satellite together are an indication of how strong the drag force is acting on the satellite. A larger size leads to a higher drag force. On the other hand, a heavier satellites leads to a lower drag force (the satellite is less affected by the drag if it is heavier). So in order to force a satellite to reenter quicker, we would like it to be as large as possible and as light as possible. Unfortunately, satellites are typically designed to be the opposite. Satellites need to be small to fit inside the rocket, but thankfully cannot be heavier than what the rocket is capable of putting into orbit. The mass over area ratio is known as the 'ballistic coefficient'. Typical ballistic coefficients for satellites are 50 kg/m² to 200 kg/m². For example, a satellite that weighs 900 kg and has an area of three by three meters (or nine square meters) the ballistic coefficient is 900 / 9 = 100 kg/m². A low ballistic coefficient will reenter a satellite quicker.

The last variable that influences the drag force is time. The reason for this: our Sun is a dynamic star. The Sun's activity moves in cycles of typically 11 years. At times of a solar maximum, the atmosphere of the Earth is heated up by the Sun and expands. The result of this is that at the same altitude, we now have a thicker atmospheric density than before, causing a larger drag forces on to satellites. At solar minima, the atmosphere becomes thinner, and satellites now reenter more slowly. Therefore, in terms of moving satellites out of the protected zones, a solar maximum is preferred. However satellites that are still operational need more fuel to keep their altitude constant, so they prefer a solar minimum. Sun spots are fairly good indications of the Sun's activity: more Sun spots mean a higher activity and more chance of solar storms. At the time of writing this book, we are in a solar maximum of solar cycle '24' as this is the 24th cycle since humans started measuring solar activity, see Figure 28 below.

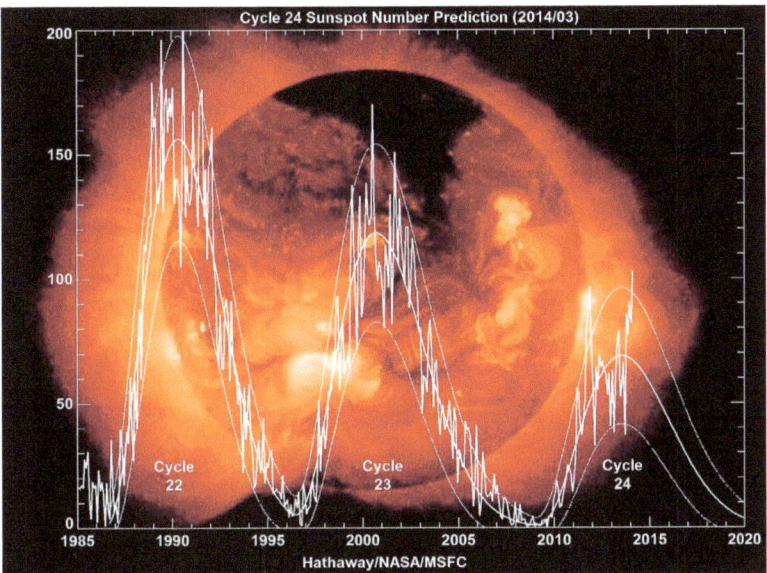

Figure 28: Number of Sun spots during the last three Solar cycles. Credits: Hathaway/NASA/MSFC.

Note that even on a daily basis the solar activity can vary. This makes it extremely difficult for satellite operators to predict when satellites will reenter. With the satellite shapes being fairly similar, the two variables: Solar intensity and ballistic coefficient are the strongest indicators of when a satellite will reenter. If we are to comply with the 25-year rule, we will need to understand this in order to understand when a satellite will reenter, and what to do if it does not reenter within 25 years. Figure 29 shows the amount of predicted time a satellite will stay in orbit and we see why for satellites with altitude higher than 600 km there is a problem. Satellites with high ballistic coefficient will already stay in orbit for almost 40 years when they are placed in a 600 km orbit. For higher altitudes, such as 700 km or higher, satellites will not reenter within 25 years even with a low ballistic coefficient of 50 kg/m². There is good news too: for satellites up to 550 km altitude there will always be a reentry within 25 years for this range of ballistic coefficients. However for higher altitudes, the only way to ensure a reentry within 25 years would be to lower the ballistic coefficient even more. This can be done with a drag augmentation device.

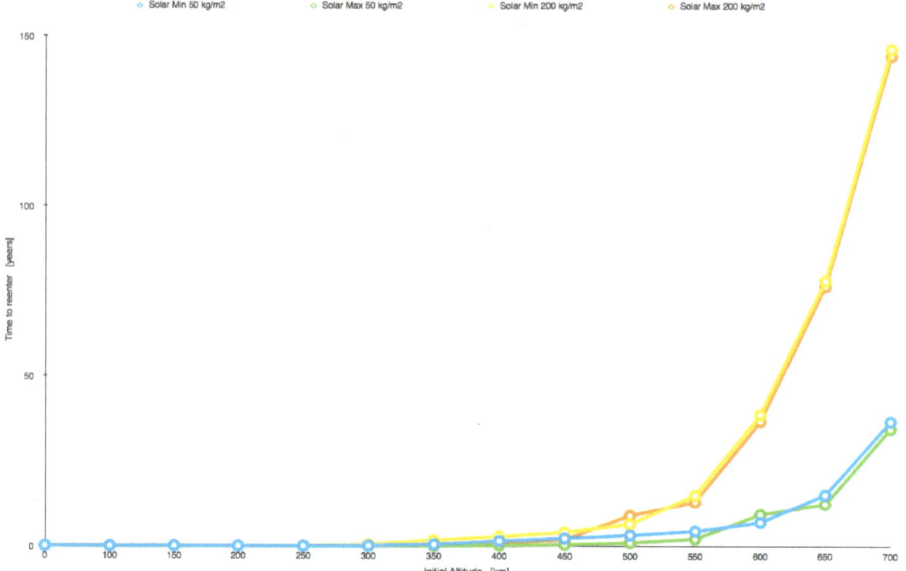

Figure 29: Lifetime of satellites as function of altitude, for Solar maximum and Solar minimum, and for satellites with low ballistic coefficient (bottom two lines) and high ballistic coefficient (top two lines).

DRAG AUGMENTATION DEVICES

Drag augmentation devices are instruments to enhance the drag force by means of increasing the size of the spacecraft without an increase in mass. In other words: to decrease the ballistic coefficient. On Earth, a parachute is a typical example of a drag augmentation device. In space, something similar can be applied, though since the air is so thin, a structure is required to keep the parachute unfolded. What we need here is a 'sail'. The figure below shows the required ballistic coefficient to reenter within 25 years, as function of altitude.

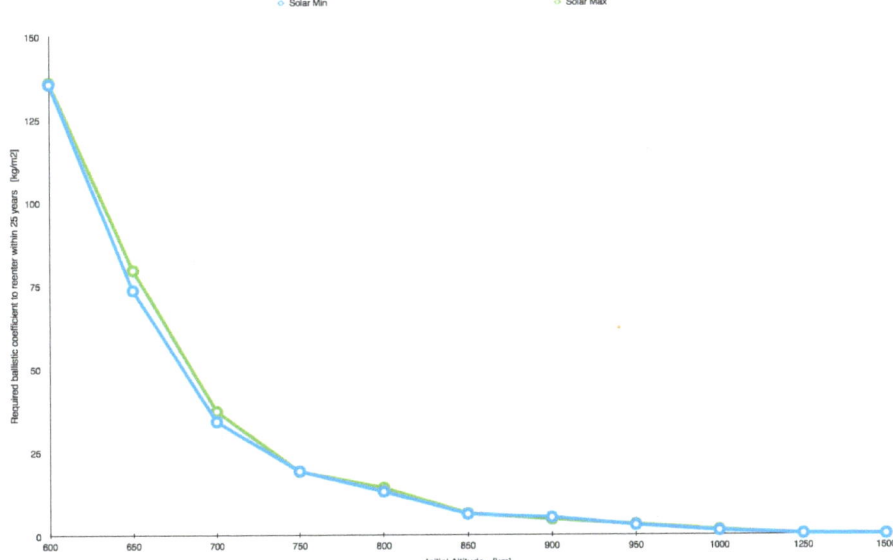

Figure 30:   Required ballistic coefficient to reenter within 25 years, as function of altitude

For example: if we have a satellite that weighs 200 kg and measures 1 x 1 x 1 meter. A side area is then 1 m². This gives a ballistic coefficient of 200 / 1 = 200 kg/m². If this satellite is placed at an altitude of 700 km, it takes over 140 years to reenter according to Figure 29. In order for it to reenter within 25 years, we need to lower the ballistic coefficient from 200 to about 35 kg/m², which is a factor 5.7. Assuming that the sail does not add much to the mass of the satellite, this means that we need to increase the size of the satellite from 1 m² to 5.7 m². In that case we reach a ballistic coefficient of 200 / 5.7 = 35 kg/m². We can achieve this by adding a drag augmentation device consisting of a sail with a deployment mechanism. During the nominal mission operations the sail is stowed ('folded' like a parachute) and at the end of the mission the sail is deployed using a few booms that deploy outward and span the sail. The sail needs to be 2.4 m x 2.4 m wide. This will give an area of 2.4 x 2.4 = 5.76 m², enough to reach the required ballistic coefficient. Figure 31 shows an example of using a sail as a drag augmentation device.

Figure 31: Sail deployed in space. Credits: Surrey Space Center.

The booms to deploy the sail could be mechanical or inflatable.

An alternative method is to deploy a long tether. If we need 5.7 m², instead of deploying a sail of 2.4 x 2.4 m, we could also deploy a 5.7 km tether of 1 mm wide. This gives the same area (5700 m x 0.001 m = 5.7 m²). A tether can be made more compact to store than a sail but the deployment is not easy. The tether needs to be unreeled or 'shot' down (or up) from the satellite in order to deploy it, and a brake will be required to stop the tether from deploying. Also, an end-mass is typically required. This makes it easier to deploy the tether, since the end-mass can be shot away from the satellite, and due to the difference in gravity forces between the top and the beginning of the tether, the tether will remain under tension.

However what if a tether is hit by small space debris? Or a micro-meteorite? In this case the tether will be cut and while one part will stay attached to the satellite, the other part will become new space debris with a relatively large size. Ways to solve this are to use tape instead of a 1 mm tether, or multiple tethers. This does however, complicates deployment and increase the mass of the system.

If a sail is punctured by a debris or micro-meteorite, it will most likely just create a small hole in the sail, but the sail will continue to function. However a sail has a problem of its own: in order to be effective, the sail should be deployed perpendicular to the velocity. With a parachute on Earth, provided that it deploys well, this is easy as there is a large drag forces and there are ropes holding the sail in the correct direction. In space, it is not so straightforward to keep the sail in the right direction. The drag forces at high altitudes are very low, due to the very thin atmosphere, and the spacecraft may have suffered a failure at its end of life while still not being in the right direction. Preferably we would need a system that keeps the satellite stable, but when the satellite reaches end of life, all systems will stop functioning. The sail needs to have a passive system to stabilize itself in a way that the drag is maximized. We can make the total system (sail + satellite) act like a shuttlecock by deploying the sail under a dihedral angle. However more sail is needed to achieve the same drag force.

A third method to augment the drag effect is to 'hit' the debris with a laser, either from ground or from space. A concentrated radar beam can create a photon pressure on the satellite, in the direction as the drag. High energy lasers would not be preferable due to the dual use as a weapon, and high costs, but low energy lasers could change the altitude of debris by a few kilometers per day. Unfortunately this method only applies to very small pieces of space debris, in the order of a few centimeters. Satellites of several hundred kilograms would need very high energy lasers.

PROPULSIVE DEVICES

Another way for small satellites in LEO to comply with the 25-year rule is to use a device that propels it out of the protected zones. Many satellites already have a propulsion system on board, for example to perform orbit correction maneuvers, or Collision Avoidance Maneuvers. In this case, the designers of the satellite need to accommodate more fuel in order to perform the de- or re-orbit maneuver at the end of life.

Imagine that a satellite is orbiting Earth at an altitude of 700 km, and this satellite has a ballistic coefficient of 200 kg/m². If we look at the trend of Figure 29, it is clear that it will take 140 years to reenter. What we need to do here is to lower the altitude to about 570 km; in this case it will take another 25 years to reenter. For small satellites that do not pose a casualty risk of more than 1 in 10,000, we only need to lower the altitude. We can then let the satellite drift down by the drag and let it reenter in 25 years. Of course, it would be better if we remove the satellite immediately out of the protected zone, for example by moving it up to an altitude above 2000 km, or even better down into the atmosphere to burn up immediately. However this will cost much fuel and as every kg put into space is very costly, satellite operators will choose for the cheaper option to lower it such that it reenters within 25 years. Lowering the orbit from 800 km to 570 km for example will cost about 5% of the satellite's mass in terms of fuel. This means that a 500 kg satellite will need to carry 25 kg of fuel to deorbit.

What if there is no on-board propulsion system already? Then the satellite designers have three options: 1) add a small propulsion system, 2) use a passive system such as the sail described in the previous chapter, or 3) add an electrodynamic tether.

Adding a propulsion system is typically not very cost effective. They are expensive systems, in particular if liquid fuels are used. Small solid rocket motor kits could be a solution as they are simple in design (similar to fireworks: you ignite them and they fire!) but the addition of a drag augmentation sail could be cheaper.

An electrodynamic tether is basically a conducting wire. By pumping a current through the tether, the interaction with the Earth's magnetic field causes a Lorenz force to be generated in the tether. With the direction of the current we can steer the direction of this Lorenz force, and we can aim it opposite to the velocity of the satellite, therefore acting as a drag force which decreases the velocity.

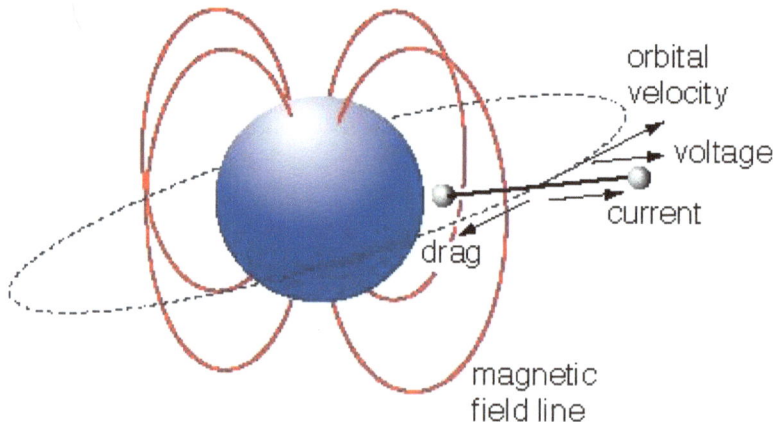

Figure 32: Lorentz force acting as drag force, created by a current through a tether within the Earth's magnetic field. Credits: NASA.

With a strong current we can even obtain much higher Lorenz forces than the drag force, and satellites could theoretically be brought down into the atmosphere within a few months, which would be a big advantage. There are three major disadvantages though: 1) the possibility that the tether gets cut by debris or micrometeorites (as discussed in the previous section), 2) the fact that we need to power the electrodynamic tether, meaning that the satellite needs to be kept alive for a few months more for it to reenter, and 3) the fact that little experience exist today with this technique. Some experiments have been done, but not always successful such as the TSS-1R experiment on board the Space Shuttle (see Figure 33). The tether experienced a peak current and broke in two. This left a 19.7 km tether as space debris in orbit, but due to the low altitude and high area of the tether, the tether reentered within a number of weeks. With space propulsion systems on the other hand, there is a lot experience, making this the most preferred solution for many satellite operators.

Figure 33: Tethered Satellite System (TSS) deployment from the Space Shuttle. Credits: NASA.

LARGE SATELLITES

Even if they comply with the rule to be removed from the protected region within 25 years, large satellites will violate the 1 in 10,000 casualty risk rule when they reenter. Designers of such satellites have two options: 1) implement a propulsive device to re-orbit the satellite up to an altitude of over 2000 km, 2) implement a propulsive device to reenter the satellite over an uninhabited zone, or 3) use Design for Demise in order to enhance the satellite's break-up during the reentry. The latter option is may not always be possible (for example when large carbon-fiber antennas need to be deployed) so here we will focus on the first two.

In order to re-orbit, a large amount of fuel is required. To ensure that effects of the Moon's and Sun's gravity forces do not influence the final orbit for at least 100 years, a final altitude of 2100 km can be chosen. This is well above the 2000 km protected zone and makes sure that within a hundred years the satellite does not decay to an altitude below 2000 km. However raising the altitude, for example from 800 km to 2100 km, costs the satellite about 25% of its total mass in terms of fuel. This is a large proportion, much larger than to deorbit it within 25 years. There are more efficient ways of doing this: a propulsion system that is based on accelerating charged ions could decrease the fuel percentage to only 1.5% compared to propulsion systems based on chemical reactions. However, these electrical propulsion based systems will need years to increase the altitude to 2100 km, while conventional propulsion systems will do this within a few hours. Adding years of operations time, after the planned lifetime of the satellite, is costly and therefore not popular with satellite operators.

Performing a maneuver to reenter into the atmosphere costs less fuel than to raise the altitude to 2100 km. For example, from 800 km it costs about 8% of fuel to place the satellite into an orbit that enters the atmosphere. This is much less than the 25% required to re-orbit. Also, it is the way to clean up space, rather than to moving the problem to another location in space. However, as large satellites typically violate the 1 in 10,000 casualty risk rule, we must ensure that the reentry takes place over an uninhabited area. An example of this is the SPOUA (South Pacific Ocean Uninhabited Area). This is a zone within the Pacific Ocean of 7000 km in longitude and 3000 km in latitude. ESA's ATV vehicles were using a reentry trajectory that ensured that the reentry would take place within the SPOUA; (see Figure 34) this we refer to as a 'controlled reentry' as we control the location where it enters the atmosphere. An uncontrolled reentry (for example in 25 years) would give a larger casualty risk than 1 in 10,000 since the ATV's weighed about 20 tons. However by using a controlled reentry and a reliable propulsion system the casualty risk was lowered to below 1 in 10,000.

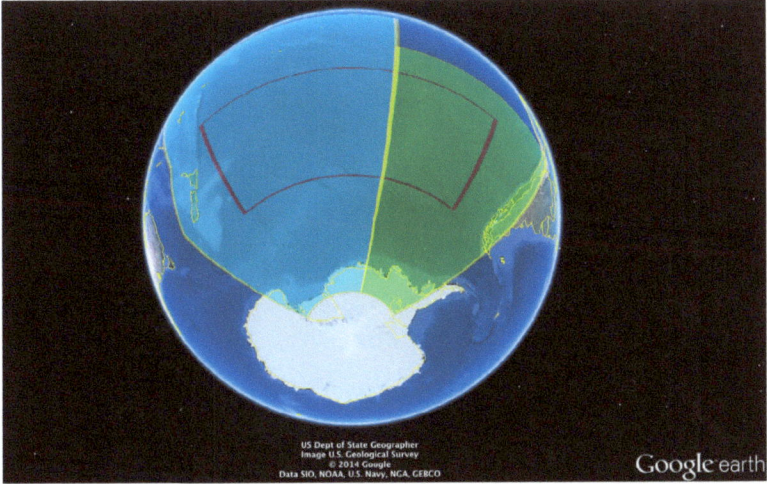

Figure 34:   Location of the SPOUA. Source: Google Earth 2014 [RD12].

The probability does not just depend on the reentry area, but also on the reliability of the satellite. For example, if the satellite fires its rockets and they are cut off half-way through the burn, the satellite would miss the reentry zone and reenter somewhere else. Many redundant and highly reliable parts will need to be put into the spacecraft to ensure the proper functioning of the deorbit function. And do not forget that this takes place at the end of the satellite's life. This means that the satellite will have spent months, even years in space before the reentry function starts, requiring high reliability even after years of not being used.

Very few large satellites are doing a controlled reentry. Apart from the ATV missions, the MIR space station and vehicles designed to survive the reentry, most large satellites do not have a function to perform a controlled reentry, or to perform a deorbit maneuver at all.

A space garbage truck would also need to comply with the 1 in 10,000 rule and will therefore need to not only catch a large debris object but also remove the debris from the protected zone and perform a controlled reentry, but also remove itself from the protected zone when its mission is fulfilled. It was mentioned before that about 8% of the total mass is required in terms of fuel. So if the garbage truck is removing 8000 kg of debris, at least 640 kg of fuel is needed. On top of this, the vehicle needs to have a structure mass, capture mechanisms such as a robotic arm, computers, fuel tanks etc. which would bring the vehicle to at least 1000 kg of mass. For this mass, a controlled reentry is required if a reentry is to be performed.

CHAPTER 4:   CATCHING A SATELLITE

Now that we have established what the function is of a space garbage truck, let's look into the methodology of catching a satellite. This is something that has rarely been done in space, and when it was done it was normally done with the help of humans on board (such as the Space Shuttle). There are several ways we can catch a satellite, and these different ways are described in this chapter. Before we do this, let's create two definitions that we will use in the remainder of this book:

- Chaser: this is the space garbage truck; the satellite that 'chases' the debris
- Target: this is the debris that the chaser needs to catch

ROBOT ARM

Looking at garbage trucks on Earth, often the garbage is stored in containers. These containers are then picked up by a large hook and emptied inside the truck. The arm is actually a kind of robot arm that has the function to catch the garbage container and to move it.

In space we can apply a similar principle: we can bring the chaser close to the target, and then use a robot arm to catch the debris. Robot arms have been used before in space (see Figure 35) but not extensively. There are several aspects that need to be considered when selecting a robot arm as the method to catch a target in space:

- The target may be rotating or tumbling around
- The chaser needs to get close to the target
- Catching a target in space is not the same as catching a target on Earth
- A robot arm contains many mechanisms.

ISS011E11416

Figure 35: Canadian robot arm on the ISS. Credits: NASA.

When an object in space is classified as space debris, its equipment to stabilize itself will not work anymore. This means that the debris could be rotating. At this point in time the motion of space debris is still unclear. There are theories that suggest that long objects could stabilize themselves automatically due to the gravity forces. The CERISE spacecraft (shown in Figure 5) used this principle during its nominal mission. If the gravity boom had not be hit by debris, the satellite would continue to stay in a stable position. There are other theories though that suggests that some debris could actually spin up, due to the interaction with the Earth's magnetic field. In any case, in contrast to on Earth, we should not assume that a target in space is stable. Even the slightest touch by the robot arm may cause the target to rotate. What we therefore need to do is to try and match the rotating motion of the target either using the chaser or using the arm itself. Figure 36 and Figure 37 explain these principles.

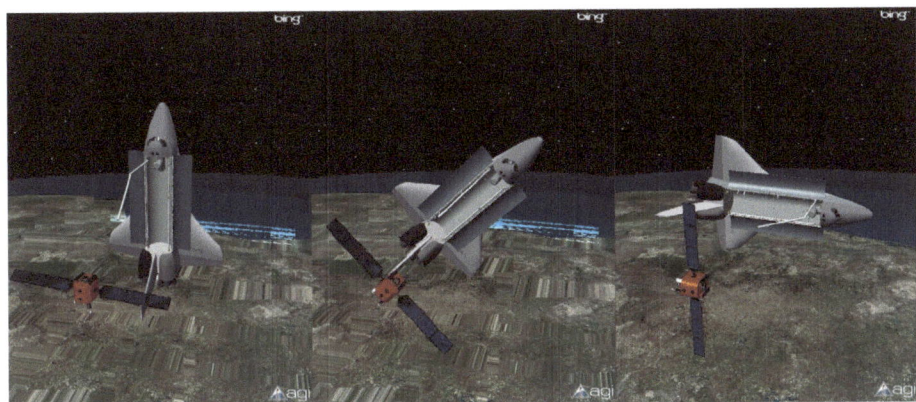

Figure 36: Example of the chaser (shuttle) trying to match the rotating motion of the target (rotating clockwise) [RD8].

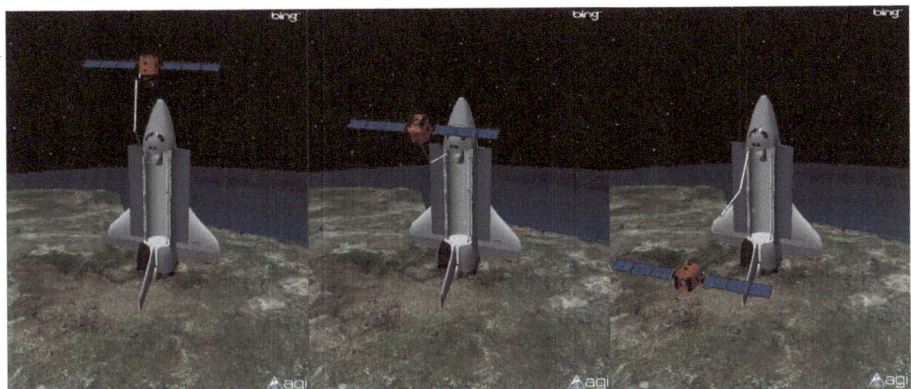

Figure 37: Example of the robot arm on the shuttle trying to match the rotating motion of the target (rotating towards the viewer) [RD8].

When the chaser tries to match the rotating motion of the target (meaning, it will be close to the target and it will look like the target is not moving, as seen from the chaser) it is called a 'forced motion'. The chaser forces itself to follow the target. Since the target appears to be standing still as seen from the chaser, it makes it easy for the arm to find a good point to grab. However the forced motion costs a lot of fuel as the chaser needs to be continuously moving and firing its small rockets. This is therefore a maneuver that the chaser cannot do for a long time, as it will run out of fuel if it does. Apart from this it is a critical maneuver: at all times guidance sensors need to keep the target in view, and perform many calculations to determine precisely how the target is moving, in order to keep following it.

An alternative is to keep the chaser satellite still, and to let the robot arm follow the target. This is more similar to what happens on Earth: the garbage truck normally parks next to the garbage container, and the truck driver uses a little camera in the arm to fine-position the arm next to the container so it can grab the container and lift it up. In space though, the target may be rotating by a few degrees every second, so the arm will need to cover a large angle if it needs time to catch the debris. For example, if a target rotates 5 degrees per second and the arm needs 30 seconds to catch the target, it needs to cover a range of 150 degrees if it is to follow the target over that period of time. In order to do so, the arm may need many 'joints' which are mechanisms functioning as elbows like in our arms. If you would try to imitate catching something on a rotating wheel in front of you, you would most likely use ('rotate') your shoulder, elbow, and wrist to make your hand follow a point on the rotating wheel. So a minimum of three rotating joints on the arm seem necessary. However the shoulders, elbows and wrists of humans are very clever things because they do not just rotate over one axis: your can rotate them in a spherical way: up and down, left and right, and combinations of that. Making two-dimensional rotating joints like that in space is very complex and expensive, so in space often a combination of two one-dimensional rotating joints are used. For example: one joint for the up and down movement, and one joint for the left and right movement.

This would bring the number of required rotating joints to six if we are to simulate a human arm, but the story does not end here: we also need a hand. In space this is called a 'gripper' as it grips the target. This is a complex thing in itself. A human hand has many muscles and many joints that allow us to place our fingers in many different positions and catch almost anything we want with our hands. However in space we only need to capture one thing: space debris, so we do not need all these rotations and we should find out what we exactly need the gripper to do.

Satellite designers however typically like to minimize the number of rotating joints in space. Mechanical parts, similar to cars on Earth or practically anything that moves, need maintenance. In space, maintenance is pretty impossible to do. It costs millions to put a satellite into space, and one cannot just simply send up a maintenance satellite to repair a debris chaser with a locked rotating joint. Moreover, lubricating mechanisms with some form of oil is not the same in space either. The mechanisms is subject to high temperature differences (for example from -200 degrees Celsius to +200 degrees Celsius every 90 minutes), and without the high gravity we have on Earth the oil may actually move away from the parts that need it. All in all, mechanisms are a risky thing in space and the thorough testing and high required reliability makes them expensive too. To develop a hand with many rotating joints will cost millions.

Another disadvantage is that the chaser needs to get very close to the target. Robot arms cannot be too long; otherwise they do not fit inside the rocket's fairing. So an arm of a few meters long is achievable, but this means that the chaser needs to approach the target to a distance of a few meters. This has the disadvantage that sensors and antennas may be blocked by the target coming into view. For example, if the antenna on the chaser that is used to communicate with Earth suddenly has a large space debris object in between the antenna and the ground station, no communication is possible. Also, lasers that are used to determine the distance between the chaser and the target may not work anymore at short distance, and cameras that are used to find the target from far distance will now have their fields of view completely covered by the target. Zooming out would be possible but this would again require a mechanism, so typically satellite designers would choose two cameras: one for long distance and one for short distance, bringing extra costs.

However one of the largest issues is being in space itself. It is hard to imagine for us humans who have not been in space or in parabolic flights what it is really like to be in weightlessness. The slightest touch makes things move, since they are not stuck to the ground. Imagine giving a target a small 'push' by accident using the robot arm or by colliding into the target and the target gets an acceleration of just 1 cm/second over a period of just one second. One hour later, the target will move away from the chase with a difference in velocity of 130 km/hour. Astronauts on board the Space Shuttle flight STS-49 tried to capture an Intelsat satellite using a clamping mechanism. However each time one astronaut touched Intelsat using the clamping device, it moved away. In the end it required three astronauts, of which two were holding Intelsat by hand, to capture Intelsat and maintain the physical connection, during one of the longest duration space walks ever done until that date.

Figure 38: Space Shuttle Astronauts capturing Intelsat. Credits: NASA.

A robot arm therefore needs to be 'clever' and respond to sudden changes in the motion of the target, even when induced by the robot arm itself. Clever means that there should be many sensors in the gripper, to image the touching point, and 'touch sensors' to understand if there is actually a touch. An intelligent computer program needs to run on board the chaser that transforms the images of the touching point into an accurate position vector relative to the gripper, so that the arm knows where to guide the gripper to. When humans grab something that is thrown at them (or is simply falling) they watch it with their eyes, which gives a three-dimensional estimate of the position of the thing that is to be grabbed (because we have two eyes), and then our brains actuate many muscles that bring our hands to the target. At the same time, our brains calculate where the target will be at time of grabbing it, and we open our fingers before touching it, and just at the right time we close them. This is a sequence of many actions that are triggered by our brains, and a robot arm will need to simulate these actions. This requires creating a very complex grabbing software which is to be stored on the on-board computer.

However if it works, a robot arm is a very intelligent mechanism that can be guided with high precision. Apart from this, once the target is captured it can even be moved in a favorable position for the deorbit maneuver.

CLAMPING MECHANISM

How can we ensure that the target does not move away when we touch it? One way is to embrace it first. 'Capture before touch' it is called. Capturing something does not imply that we touch it. In the end we do want a firm grip on the target, so that we can move it, but before we touch the target, it would be a benefit if it is captured already. A kind of tentacle would help here in order to embrace the target first. Since the word 'tentacle' may give an image of a arm with many rotating joints which is not desirable as we discussed in the previous section, we will refer to this as a 'clamping mechanism'. Figure 39 gives an example of a clamping mechanism.

Figure 39: Example of a chaser using a clamping mechanism to capture a target. Credits: ESA.

When selecting a clamping mechanism to capture a target in space, the designers should take into account the following aspects:

- Clamping mechanisms may be very long if the target is big
- We must find a way to ensure that after capture the target does not move around
- It may be difficult to accurately position the chaser on to the target

One advantage of using a clamping mechanism is that we can reduce the number of rotating mechanisms compared to a robotic arm. In the example shown above (Figure 39) there is only one rotating mechanism per clamp, and the total number of clamps is four. However this means that we need long arms of the clamping mechanisms, and they should be far apart in order to completely cover the cross section of the target satellite. For example, if the target satellite has a cross section of 2 x 4 meters, the clamping mechanisms should be four meters long and two meters apart. Apart from needing long arms, we also need a wide chaser. This could pose problems fitting the chaser into the rocket that launches the chaser. Small rockets have an available diameter between 2 and 3 meters, so the designers of the chaser need to make sure that everything fits. The long clamping mechanisms may need to be stowed vertically but even four meters could give problems with fitting inside small rockets.

When the tips of the clamping arms are folded, as is shown in Figure 39, we can embrace the satellite and make sure that the target cannot escape when the clamps close. However, there will still be some freedom for the target to move within the closing boundaries. In order to perform an accurate controlled reentry we must make sure that the target cannot move around: we will need to fix the target and for this there we cannot escape using another mechanism. One way would be to add a mechanism that pushes the target against the clamps. Another way is to pull back the clamping mechanisms, in order to pull the target against the chaser using the clamping mechanisms. Figure 40 below gives an example of pushing the target against the clamps, using pushing rods.

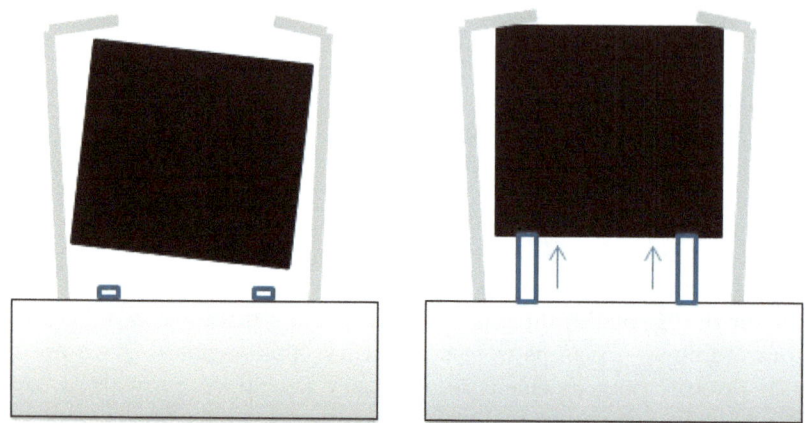

Figure 40: The left side shows that the target (black box) is captured by the grey chaser, but still has freedom to move. The right side shows how pushing rods are fixing the target.

Unfortunately, this fixation will take a bit of time. We cannot simply shoot a fixation device with high speed against the target, as it may break. Also, the mechanism will have a kind of gearbox that can exert a larger force if the mechanism rotates more slowly (similar to a first gear in a car compared to a fifth gear). The problem with this is that within the few seconds that the fixation device (like a pushing rod) needs to fix the target, the target may move within the clamps. If it takes 5 seconds to achieve a strong fix and the target still has a bit of relative velocity (which could have easily been exerted by one of the clamping mechanisms when it touched the target), for example 1 cm/second, then the target could move 5 cm sideways within the time it takes to fix it. It will therefore be very difficult to accurately position the chaser on to the target. Most likely an accuracy of a few cm cannot be achieved. And in space, we may need a better accuracy if we are to perform a precise controlled reentry, as will be explained in the next chapter.

NET

Fishermen have used fishing nets for over 5000 years so the use of a net is a proven method to catch a moving target. So why not use this method in space? A net would be able to fully encapsulate the target, which has several advantages.

A fully encapsulating net would ensure that if any large pieces breaks off from the target, it stays inside the net (and does not become a new space debris object). Also, there are many places where the net touches the debris which means that when a force is applied, this force is nicely distributed over a large surface of the target. A robot arm would flow the entire force through the small touching point with the target, giving a higher chance of breakage at that point.

A net is also light and can be folded into a small volume. The shape of the target is of little importance: a net can easily catch a cube form (like a satellite) or a cylinder form (like a rocket upper stage). And when a net containing a target is pulled by the chaser using a tether, it does not matter what the orientation of the target is. The net is pulled into one direction only, which will have a stabilizing effect on the target.

One very important advantage is that the chaser does not need to get close to the target. For a rigid connection the chaser needs to touch the target, and therefore during the rendezvous they may collide due to appendices sticking out of the target, such as antennas. For a capture based on a flexible link such as a net, we can shoot the net from a safe distance.

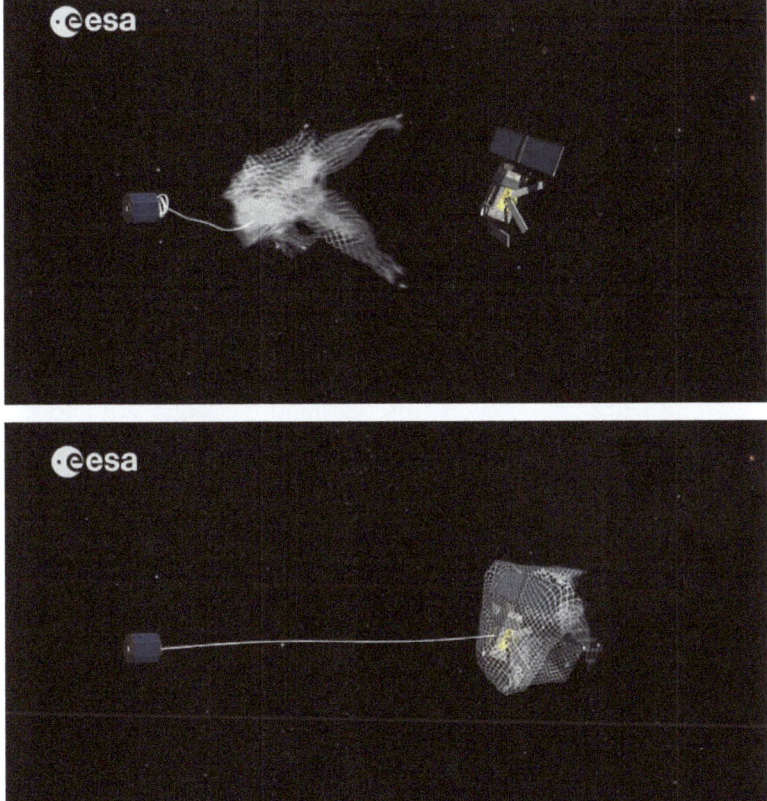

Figure 41: Example of a chaser using a net to capture a target. The top image shows the deployment. The bottom image shows the target captured. Credits: ESA.

Still, there are many design considerations to be considered:

- If the target is heavy or rotating fast, the tether may wind around the target
- It is difficult to have several attempts as each attempt requires a separate net
- When the chaser does the deorbit burn, it fires directly towards the tether and the debris
- It is complex to control tethers in space.

Fishermen normally throw a net over the fish; a special technique which needs to be learned. To shoot a net from a chaser spacecraft, we need to find a different technique. The most obvious solution is to shoot a number (for example, four) bullets to which a square net is attached at each corner. However how do we close the net around the target? The net might close itself but entangling around antennas and solar panels of the target. However to be sure we could implement winding mechanisms in the four bullets, which wind up a rope in between each bullet. This entire process will need to be automated: the net needs to be shot from a certain safe distance, for example 20 m, at a certain velocity, say 2 m/second. Then it takes 10 seconds for the net to hit the target, and timers within the four end bullets will then activate the winding mechanisms.

Since the net does not put any force on to the target, the target will keep rotating. This creates the problem that the target may wind up the tether that spans from the chaser to the target, causing the chaser to eventually hit the target. The chaser will need to pull from the tether at the right time in order to dampen the rotating motion of the tether. However this can only be done once it is confirmed that the net is closed around the target. And how could a computer program inside the chaser know that the net has closed around the target? This is a technical difficulty in itself. One solution could be to use another timer, but it is not guaranteed that the net is actually closed within the given time.

What happens if the capture goes wrong? If we miss the target entirely, we have created yet another space debris object in the form of a large net. If we hit the target, but somehow the net does not close properly, we have another problem. In either case, we would need to cut the tether and with it the net, and leave it where it is. It is simply impossible to reel up the tether and stow the net in a nicely folded way, and re-arm the mechanisms to shoot the end bullets away. If we want to re-attempt a capture, we need to bring another net (with its own shooting mechanism). Since mass and available volume is limited on a satellite, we cannot bring a large number of nets. Possibly only two or three, as nets big enough to enclose entire satellite will be very big, and will take quite some space even when stowed away.

If we look at Figure 40 above, we see the chaser on the left side and the target captured by a net on the right side. To deorbit the target, the chaser needs to pull the target to the left side of the picture. This means firing its thrusters towards the right side. However the tether is also located to the right side of the chaser, which means that the thrusters will be throwing their flames in the direction of the tether. The designers will have to be careful not to burn the tether so that it breaks in two. One way is to off-point the thrusters (meaning that they point a bit away from the tether) but then the thrusters are not exactly pointing into the right direction. This will make them less effective and in turn more fuel is needed to perform the deorbit burn. Another way is to shield the tether. For example using a TPS which is normally quite expensive, or selecting a material for the tether (or part of the tether) that is heat resistant, this may be quite heavy.

However one of the largest problems is the behavior of the chaser and the target when they are connected by a tether. As mentioned in the previous chapter there is still little experience in using tethers in space. Typically wires have some elasticity, which means that following a deorbit burn the stretched tether may start pulling the chaser and target towards each other. A deorbit burn could consist of several burns, so we need to make sure that the system is controlled in between the deorbit burns. This means keeping the tether under tension to avoid it from going slack in space. And this requires a complex control system and possibly continuous firing of small thrusters. Gemini 11 astronauts attempted to keep a tether under tension while creating artificial gravity in space, but found it difficult to avoid the tether from getting slack as shown in Figure 42. The behavior of a large net and tether in space cannot be tested on Earth due to the presence of gravity. It can only be simulated. This may not convince stakeholders to finance the mission.

Figure 42: Tether experiment on Gemini 11. Credits: NASA.

HARPOON

Another device used by fishermen is a harpoon, and in space we can use a similar device. To penetrate the metal body of a satellite is more difficult than to penetrate the skin of a fish. However once shot into the body, the fixation point is fixed in contrast to that of a net which may move around the target. Another huge advantage is that shooting a harpoon can be tested on ground. Like the net option, the chaser does not need to touch the target.

Nevertheless, all the disadvantages that apply to the net, as described in the previous section, also apply to the harpoon. We still need a tether connecting the target and the chaser.

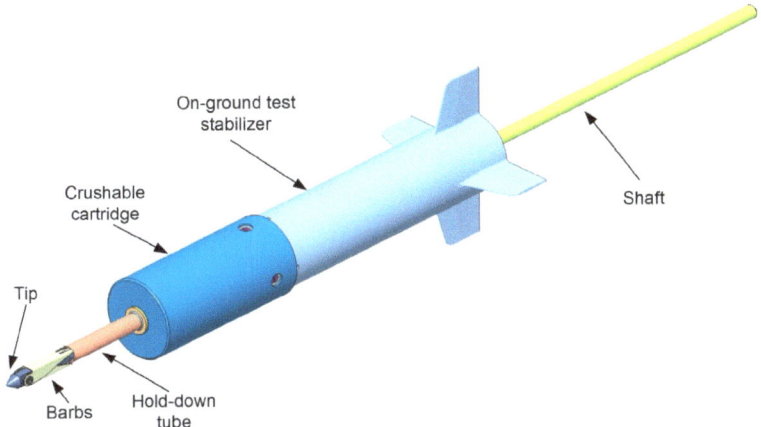

Figure 43: Example of a harpoon design to catch space debris. Credits: Airbus Defense and Space.

A potential problem with a harpoon is that we need to be careful not to hit a fuel tank or battery inside the target. In case there is still fuel on board for example, a spark could ignite the fuel and the target may explode. Another issue is that the chaser needs to get closer to the target than in the case of the net, as the harpoon needs to be shot into the target in a very accurate way. As any fishermen will be able to tell you, hitting a fish from 5 meters distance is easier than hitting a fish from 50 m distance. We will have to make sure that the harpoon is fixed after hitting the target so we will need a clever confirmation method so that the harpoon does not come loose when the chaser starts pulling the target. Finally, small pieces of debris could be released upon impact of the harpoon. Since the objective is not to release more space debris, the design should be such that most of these pieces should be ejected inside the target instead of outside.

LASSO AND OTHER METHODS

Some other methods are described here that are worth mentioning. The first is the use of a lasso. This could be achieved in two ways. The first is to wind a tether around the target. However this is a very complex task as first the tether needs to be attached to the target, and secondly the winding will take quite a bit of time and in the meantime the tether may start to move around as the motion of tethers not under tension is extremely difficult to predict.

A second implementation of a lasso could be to have a rigid ring which can be extended to fully fit the target inside the lasso. However a ring like this would be very difficult to stow within a spacecraft. To fit the target inside the ring, another rigid connection is needed between the ring and the chaser, which is another part that takes up much volume. Like the harpoon case, the chaser will need to get close to the target.

Wrapping can be applied in a similar way: a sticky material could be wound around the target, possibly using the tumbling motion of the target itself. However there is very little experience in applying sticky tapes in space and vacuum, and the problems of the lasso described before are still applicable.

Magnets would be an obvious choice to create a connection between two bodies in space. However satellites structures are nowadays often made from carbon fiber–reinforced polymers, which do not have good magnetic properties. And most satellites are covered with multi-layer insulation (MLI; a kind of gold foil) to increase the thermal properties. These layers complicate getting a good magnetic surface to magnetic surface grip. Finally, strong magnets may disturb sensors and actuators on the chaser satellite itself, which are needed to accurately maneuver itself through space.

Foam could be applied as well; foam that either encapsulates parts of the chaser or has glue-like properties. However a large amount of foam would be required as satellites are several meters big. And when a connection is found via foam, this connection would not be firm enough to withstand the large forces that the chaser applies during the deorbit burns. The deorbit burns are performed with large thrusters that put several hundred Newton of force on to the target via the connection with the chaser.

The first options: the robot arm, clamping mechanisms, net and harpoon, seem therefore a more realistic choice for a capture mechanism for space debris.

CHAPTER 5:   DESIGNING A GARBAGE TRUCK FOR SPACE

In this chapter we will take a closer look at design aspects of a space garbage truck. A satellite in space normally consists of a 'payload module' and a 'service module'. The payload module is the part of the satellite responsible for delivering the wanted result. For a scientific satellite, this would be the sensors measuring the requirement measurements. For example, infrared pictures of the Sun, or heat waves from forests on Earth. The service module is the rest of the satellite that helps getting the payload in the right position, orientation and temperature. It also provides the communication with ground stations on Earth. So the service module consists of a carrying structure, antennas, a propulsion system, gyros, heaters and/or cooling systems.

From the functional diagram in Figure 20 we know that a garbage truck on Earth has the following functions:

1. Go to the neighborhood of the garbage collection
2. Get precisely to the garbage
3. Capture the garbage
4. Move the garbage to a central place
5. Dispose or recycle garbage

In space we can define similar functions:

1. Go to the neighborhood of the debris
2. Get precisely to the debris
3. Capture the debris
4. Move the debris into the atmosphere
5. Dispose the debris

Let's take a look at which functions are defined for the payload, and which ones for the service module.

PAYLOAD MODULE DESIGN

One can argue that all parts of the satellite are responsible for the mission success. Nevertheless, there are certain functions on a space garbage truck which are very specific to its mission to clean up space. We can define 'any function related to the capture of the space debris' as payload. This is not only the function of the capture mechanism, but also any sensor (such as cameras) necessary to capture the target. For the five functions defined, functions 2 and 3 would be allocated to the payload, and the other functions to the service module.

1. Go to the neighborhood of the debris (service module)
2. Get precisely to the debris (payload)
3. Capture the debris (payload)
4. Move the debris into the atmosphere (service module)
5. Dispose the debris (service module)

Let's start with function 2; this is the first function of the payload. We are already in the neighborhood of the debris since we completed function 1, but what is 'in the neighborhood'? One way to distinguish 'far away' from 'in the neighborhood' is the point when the target starts to be visible to the chaser, i.e. when it is picked up by a camera. For this function then, a camera with a large focus (typically called Narrow Angle Camera) is required. The chaser then needs to navigate towards the target. Navigation towards a target in sight is called 'relative navigation' as the chaser will continuously compare its own position relative to that of the target, using the camera. The navigation algorithm will need to calculate how far left or right, above or underneath the target is from the current course of the chaser, and give commands to the thrusters on the service module to steer left or right, for example. However it also needs to understand the distance. Distances are often measured using a laser.

These sensors are often called 'Laser Range Finder' or 'LIDAR' (Light Detection And Radar) and measure the time it takes for light to be reflected back from the target to the chaser.

In case we need to capture the target using a robot arm and/or clamping mechanisms, we would like to 'hold still' the target as much as possible. This means that if the target is rotating or tumbling around, the chaser should follow its motion. This is a complicated task. First the chaser needs to use a camera to film the target. Then, a sophisticated algorithm in the chaser need to determine, from this movie, how the position of the target is with respect to the chaser, and how is it rotating (e.g. what is the rotation axis? Where is it pointed to?). When this is done, the chaser needs to maneuver itself around the target in order to keep one of its sides (the side where the chaser wants to grab the target!) continuously pointed towards the chaser. This is the forced motion described in the previous chapter. Alternatively, we can rotate the arm but if the target rotates fast or in a strange tumbling way, this will be difficult. The forced motion requires small thrusters to continuously fire in order to continuously correct the position of the chaser, so that it keeps a fixed position to one side of the target. If we are to remain close, the Narrow Angle Camera will only see a small part of the target. We therefore also need a Wide Angle Camera to be able to see the entire target even if we are close.

Needless to say, the chaser will still need to be able to send its data in telemetry form down to Earth, and should be able to receive telecommands. With the space environment still acting on the chaser, we can create the following IDEF functional diagram of the function to 'get close to the debris':

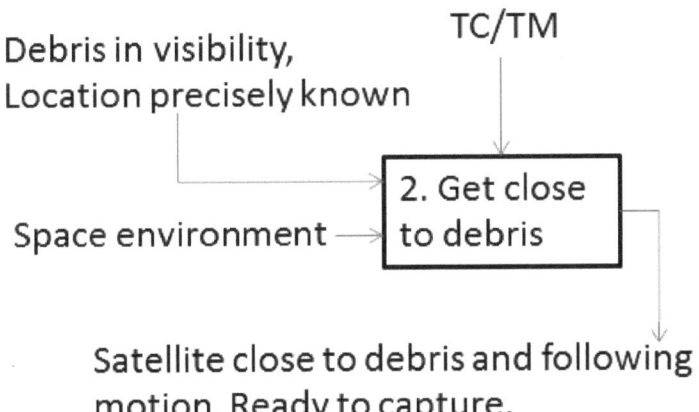

Figure 44: Functional diagram to get close to the debris.

We know that in terms of payload module we need the following equipment:

PL1: Narrow Angle Camera (NAC)
PL2: Wide Angle Camera (WAC) only if we need to get close
PL3: Laser Range Finder or LIDAR
PL4: Navigation computer with Image processing algorithms

Furthermore we will need some small thrusters to guide the chaser to the target. Since the service module will need to have (big) thrusters for the deorbiting function, it needs to be decided if the small thrusters used for navigation are part of the payload module or service module. For the sake of this exercise, let's assume that they are part of the service module to have one integrated propulsion system. The picture below show the Narrow Angle and WACs of the OSIRIS payload (Optical, Spectroscopic, and Infrared Remote Imaging System) used on ESA's Rosetta mission to take high resolution pictures of a comet.

Figure 45: NAC of ESA's Rosetta mission. Credits: MPS [RD13].

Figure 46: WAC of ESA's Rosetta mission. Credits: MPS [RD13].

Now let's have a look at function number three: 'Capture the debris'. At this point in time we are close to the debris; we have it in view and are ready to grab it. Camera images will probably be sent down and the chaser will wait for a 'go' decision from ground. When it is received, the chaser will start the capture process.

If the chaser is designed for a flexible system such as the net or the harpoon, the chaser will not need to do a forced motion. It can simply wait from a distance of a few meters (harpoon) to a few dozen meters (net) and wait for the right moment to shoot the capture device towards the target. If the target rotates in a rapid way, this timing of shooting the net or harpoon away will need to be automated, based on the images that are processed within the on board computer.

When the capture sequence is completed, a confirmation is necessary to determine that the target has actually been captured. With a robot arm this could be sensors that measure the force of the grip. With a net this could be a measurement that all the end masses have completely wound up the ropes in between them. With a harpoon we can measure if the barbs on the tip (see fig40) have been deployed. The capture confirmation is a complex measurement.

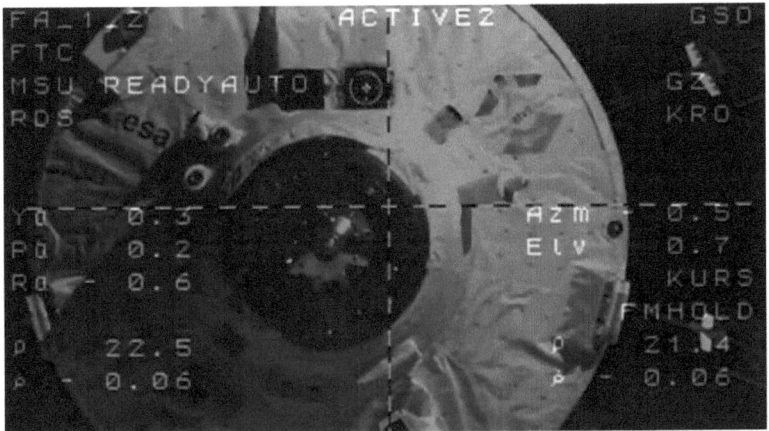

Figure 47: The Automated Transfer Vehicle 5 (ATV-5) docks to the ISS using very similar sensors required to capture debris, such as a LIDAR. Credits: ESA.

From the moment we captured the target, the target will excite forces on to the chaser via the capture system. For a flexible link, a rotating target will start pulling the tether in between the chaser and the target. For a rigid link, we now have a new 'satellite' that consists of two satellites attached to each other. The target will start rotating in a different way, since the mass of the total rotating system has now increased. If the target is much larger than the chaser, this effect will be much stronger than if the target is smaller than the chaser. If the chaser was doing a forced motion, it should now stop this forced motion. However the target will need to be stabilized, so the chaser will need to start firing its thrusters immediately in order to stabilize the target via the tether (by giving pulling forces at the right time) or to stabilize the stack of chaser and target if they are connected rigidly, This is a new type of forced motion, though this time to halt the motion rather than to follow the target. Many complex algorithms will need to run at this time. For example, a satellite (or stack of satellites) rotates around its center of gravity and the algorithm will now need to calculate what the new center of gravity of the combined target and chaser stack is.

The figure below shows the functional diagram. As inputs we have the output of the previous function (that we are close and ready to capture), the space environment and the tumbling motion of the debris. We can still receive commands (for example, to abort) and send telecommands (for example, camera pictures like the ATV-5 picture above). However once attached we also have a physical interface to the target, which is shown below the block. As output we have the state that the debris is now attached, and the function does not end until the entire chaser + target stack is stabilized.

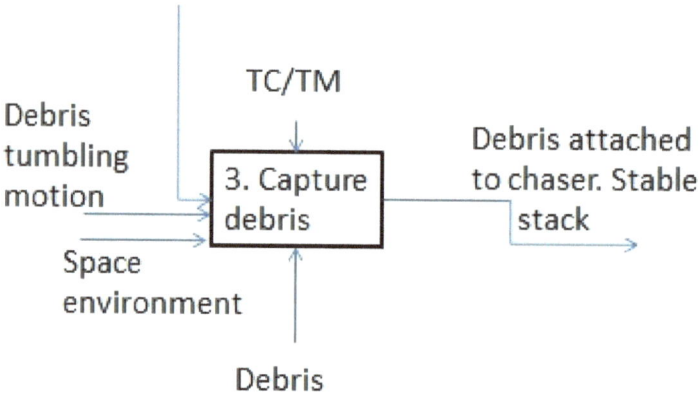

Figure 48: Functional diagram to capture debris

In terms of payload module we need the following equipment:

PL5: Capture device
PL6: Sensors for capture confirmation
PL7: System (actuators and algorithms) to stabilize the target

In terms of satellite configuration, it is clear that the cameras will need to be pointed towards the target. And the capture mechanism should be able to reach the target in the same direction. One could therefore expect that one side of the satellite is dedicated to payload module equipment (cameras, laser, net shooting mechanism or robot arm). The navigation computer could be placed inside the satellite but will most likely be placed not far from the plate to which the cameras and laser are connected, in order to avoid long electrical lines. Small thrusters will need to be put in such a way that the chaser can thrust in any direction including the direction of the target. These are part of the service module in terms of functionality, but we can see that some overlap between service module and payload module in terms of placement will be necessary.

SERVICE MODULE DESIGN

Now let's take a look at the 'machine' that ensures that the payload module can do its job properly. We can already identify two major tasks, such as getting the payload to the target, and removing the target from orbit, but there are other functions as we will see when we go more into detail.

Function number one that we identified is 'Go to the neighborhood of the debris'. However we should not forget that a space mission starts on ground. Before we can even get close to the debris, we need to get into space first using a rocket that launches the chaser. This launcher will pose strong constraints on to the design of the chaser: it needs to fit inside the volume of the launcher, it is limited in mass by how much the launcher can actually put into orbit, it needs to survive the shocks of the launch, and it needs to have a specific interface to the launcher. Launcher user manuals can normally be downloaded from the website of the launcher fabricant, for example from [RD14]. Inside the user manual we can read what shock levels the chaser needs to comply with, and how the interface works. Often launchers work with a 'Launcher Adapter Interface' which connects to the satellite via a clamp band system. When the launcher has reached the required orbit in space, the clamp band system ejects the chaser away from the upper stage of the launcher. And hereby we have defined one of the first functions of the service module: to provide a proper interface to the launcher. The launcher will transmit heavy loads via the clamp band to the satellite so the service module will need to provide a solid structure to transmit these loads and make sure the satellite does not shake into pieces. For this reason, a satellite often has a central tube with exactly the same diameter as the Launcher Adapter Interface. A launch is typically in the order of 10 to 20 minutes. During this time, the service module will need to rely on batteries in order to power the entire chaser.

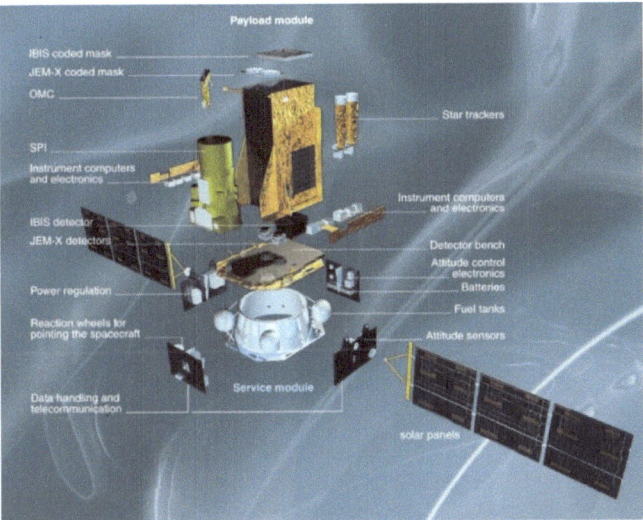

Figure 49: Example of a satellite exploded view. ESA's Integral satellite consisted of a service module and a payload module (gamma-ray telescope). Credits: ESA.

Once in space, the space environment takes over the launch environment. This means not only that the chaser is bound by the Earth's gravity, but strong thermal fluctuations will act on the spacecraft which the service module needs to handle. In cold cases (such as being in the shadow of the Earth) heaters will need to keep the instruments above their minimum required working temperature. In sunlight, heat created by the equipment will need to be radiated to space. This is called a thermal control system.

Another aspect is that the service module will need to provide power to all equipment on board. This means that from the moment it is separated from the launcher, it must make sure that sunlight falls on to the solar panels. The service module must have a power system to convert power from the solar panels and distribute this power to all equipment. It must also charge the batteries when the satellite is in sunlight, and switch to battery power when the satellite is in shadow.

If solar panels are stowed away, we must immediately deploy them. Some mechanisms will then be required to deploy these panels outwards. Then, the spacecraft needs to be turned in the right direction to maximize the Sun energy on the solar panels. This implies two things: first of all we need to know what the position of the chaser in space is, and secondly we need some way to turn the chaser. The service module will need to have sensors such as a Sun sensor, Earth sensor and possibly even a star tracker to measure its own attitude relative to the Earth, Sun and stars. It can then calculate its own position and how much (and in which direction) it will need to rotate to maximize the Sun energy on the solar panels. Rotating a satellite can be done in different ways. Gyros are often used to keep the satellite stable, and by accelerating or decelerating the gyros the satellite can rotate around the gyro. A similar effect can be obtained with reaction wheels. Another example is using small thrusters to fire on one side of the structure to make the satellite rotate, and then fire another thruster in the opposite direction to stop the rotation. Since we know from the payload module that thrusters are needed in any case to guide towards the target, we could use the same system to control the attitude. The system to measure both the attitude and position of a satellite in space, and to control it, is often referred to as Attitude and Orbit Control System (AOCS). This includes all algorithms necessary to do the attitude and orbit computations, and the computations to determine which thruster(s) to fire.

In order to talk to ground stations on Earth, meaning sending and receiving data, a communications system is required. Not only does this consist of antennas (which should be redundant in such a way that no matter the attitude of the chaser it can still transmit a signal to ground), but also of a transmitter, receiver, and modules to generate signals in the correct frequency and polarization, including error correction coding. Data needs to be created that gives the state of the chaser. This is called house-keeping data and consists of, for example, the temperatures of the equipment, information on the calculated position and attitude, which commands have been executed inside the satellite etc. Apart from house-keeping data, the data produced by the payload module, such as images by the WAC and NAC, or confirmation of successful capture, also needs to be sent to Earth. This data is generated by a Data Handling System (DHS), consisting of an on-board computer and all connections to all equipment on board. The DHS also needs to handle the incoming commands from Earth. The commands have to be interpreted, and executed within the chaser.

In summary, we can create a diagram of function one in the following way:

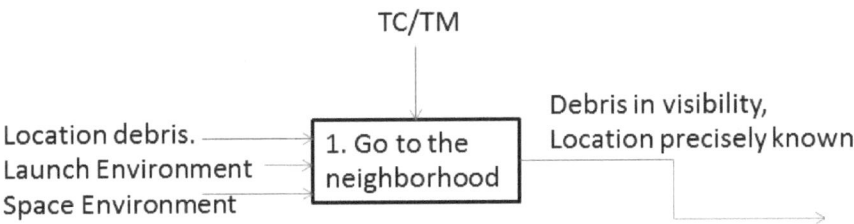

Figure 50: Functional diagram of the function to go to the neighborhood of the debris

Inputs are the location of the debris. The debris can be tracked from ground and a position can be obtained with reasonable accuracy, enough to bring the chaser into the neighborhood. Other inputs are the launch and space environment. Commands are given and received using telecommands and produced telemetry. The service module will then guide the chaser towards the target until the NAC gets the target into view, and function 2 can start. For this function we know we will need a lot of equipment:

SV1: Structure (including possible mechanisms to deploy solar panels)
SV2: Thermal control system
SV3: Power system
SV4: AOCS (including small thrusters)
SV5: Communications system
SV6: DHS

Once the payload module has completed its job and the target is captured and stabilized, it is now up to the chaser to remove the target from its orbit. Functions four and five are allocated to this. Function four is to 'move the debris out of the protected zone'. As described before, if a satellite is in LEO the chaser can bring the target into the atmosphere to burn up, or re-orbit the target to an altitude higher than 2000 km. Both cases require accurate pointing of the chaser with the target connected and a propulsion system with enough energy to move both target and chaser. During the firing of the thrusters the chaser needs to ensure that it does not lose the target, or in the case of a flexible connection, collides with it. For the case of a reentry we can create the following functional diagram.

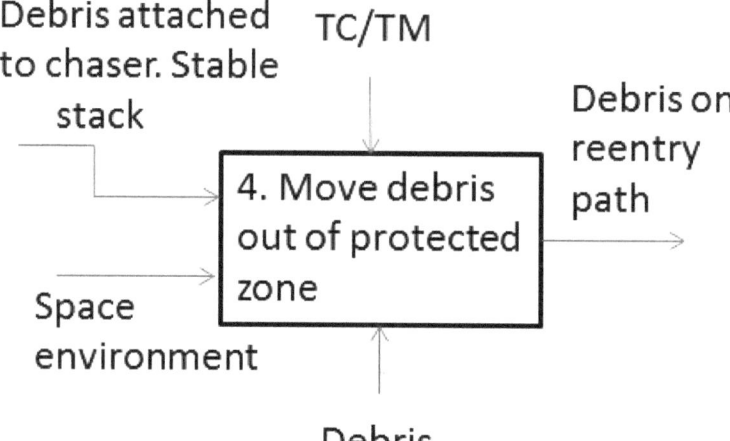

Figure 51: **Functional diagram of the function to move the debris out of the protected zone**

As extra equipment on the service module we need:

SV7: de- or re-orbit propulsion system

If we are to reenter a target, a large propulsion system is needed able to provide a high thrust. High thrust is required to accurately navigate the stack into the atmosphere. At the same time the thrust cannot be too high to give the target such a large acceleration that pieces break off. However, high thrust systems are normally not the most effective systems. As stated earlier, a deorbit burn from 800 km into the atmosphere requires a fuel mass of about 8% of the entire moved mass. So if the chaser is to remove an 8000 kg heavy target, at least 640 kg of fuel is needed to move this chunk down into the atmosphere.

In the last decades new propulsion systems have been developed, based on electrical energy. This 'electrical propulsion' has a much higher efficiency. However the thrust level is so low that it will take many months to move an 8000 kg target. Since the Earth's atmosphere may thicken or get thinner in those months, it will be impossible to perform an accurate reentry over an uninhabited area. However, these low thrust systems could be used to re-orbit a target to above 2000 km. For example, to move a target from 800 km to 2100 km, a fuel ratio of 1.5% and transfer time of two years is required using low thrust. In this case the fuel would be 120 kg instead of 640 kg for the high thrust case, at the expense of two years of extra operations (maintaining a team on ground to communicate with the chaser).

Finally, when the chaser moved the target out of the protected zone, we should dispose of the target. If we moved the target up to an altitude of more than 2000 km, we simply need to detach it from the chaser and ensure that no collision takes place. This can be done by raising the orbit of the chaser after the target has been released.

In the case of a reentry trajectory, we hope to burn up as much as possible of the target. The trajectory should be such that the footprint of possible surviving debris is as small as possible. This can be done by entering the atmosphere as steeply as possible. If we take debris mitigation regulations into account, both chaser and target shall not pose a risk of casualty on ground higher than 1:10,000. This can be achieved in several ways. First of all reentry zone over an uninhabited area is to be chosen such as the SPOUA (Figure 34). Secondly, the chaser service module should be designed such that the probability of failure is low enough to ensure this casualty risk constraint. This means adding redundancy to the deorbit function (in particular on the propulsion system). Figure 52 below shows the functional diagram.

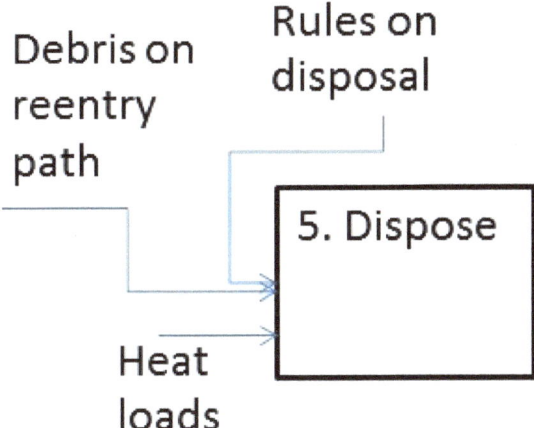

Figure 52: functional diagram of the function to dispose of the debris

The diagram shows the reentry case, where heat loads act on the chaser and target, causing both of them to disintegrate when entering into the atmosphere.

Finally, Figure 53 below shows all the functions combined, creating the functional diagram of the space garbage truck, i.e. the mission of the chaser.

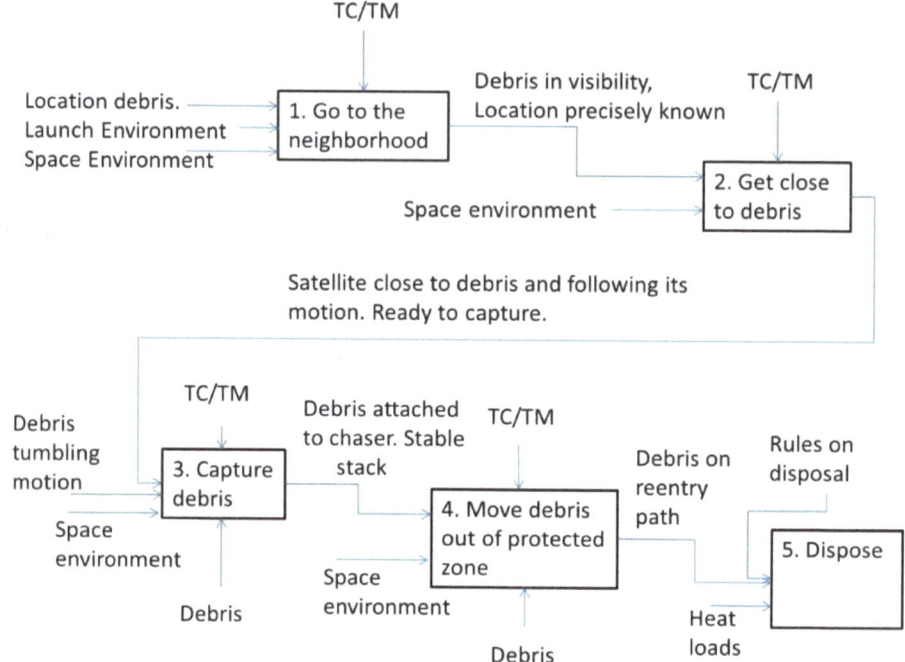

Figure 53: Functional diagram of the entire chaser

SYSTEM TRADE-OFFS AND DESIGN

Now that we know all the building blocks that form a space mission, let's use a system engineering approach to put these building blocks together and form a mission that reaches the required objectives. The design will depend on the mission objectives of the mission. Possible mission objectives are, for example:

- Deorbit a pre-selected debris satellite into the atmosphere, or
- Remove a minimum of five debris satellite from the protected zone, or
- Remove an x amount of rocket bodies, or
- Perform some service operations to a satellite and then deorbit it into a lower orbit to comply with the 25 year rule.

Each of these objectives will lead to a different design. For example, if we are to remove ten rocket bodies, it may be difficult to use the net capture technique even if it is independent of the rocket body shape. For each rocket body we need a separate net, and we need to insert redundancy. Putting at least 11 nets on a satellite, including ejection and closing systems, will be much heavier than putting one robot arm which can be reused for all rocket bodies.

And how do we divide equipment between payload module and service module? For a mission to reenter a specific satellite, we could optimize the configuration and possibly integrate parts of the payload and service module into one compact satellite. A smaller and lighter chaser could fit in a small and cheaper rocket. Small rockets can be bought for a few tens of millions of dollars. However if the chaser is to de- or re-orbit multiple targets, perhaps creating small payload modules including propulsion systems, which can be detached from the service module and attached to the target to do the removal function, is more optimal. These would be 'deorbit kits' where each target gets one kit installed. The service module could then simply be a carrier using efficient low thrust propulsion to get the payload modules to each of the target. A propulsion system based on solid propulsion (a system carrying solid fuel) could be used as it is simple and compact. This would lead to a larger and heavier chaser, requiring a heavy rocket. Unlike smaller rockets, heavy rockets cost well over a hundred million dollars. This would make the launch very expensive, however if we divide the total mission cost per debris removed, it may be cheaper per debris removal than one debris removal per launch. It all depends on the design outcome and is a trade-off that system engineers need to take, making sure that the mission objectives are achieved in the best possible way.

Figure 54 below shows a typical trade-off table that space system engineers need to create in order to help taking decisions based on desires by the project (such as mission objectives) and constraints (such as the mission budget).

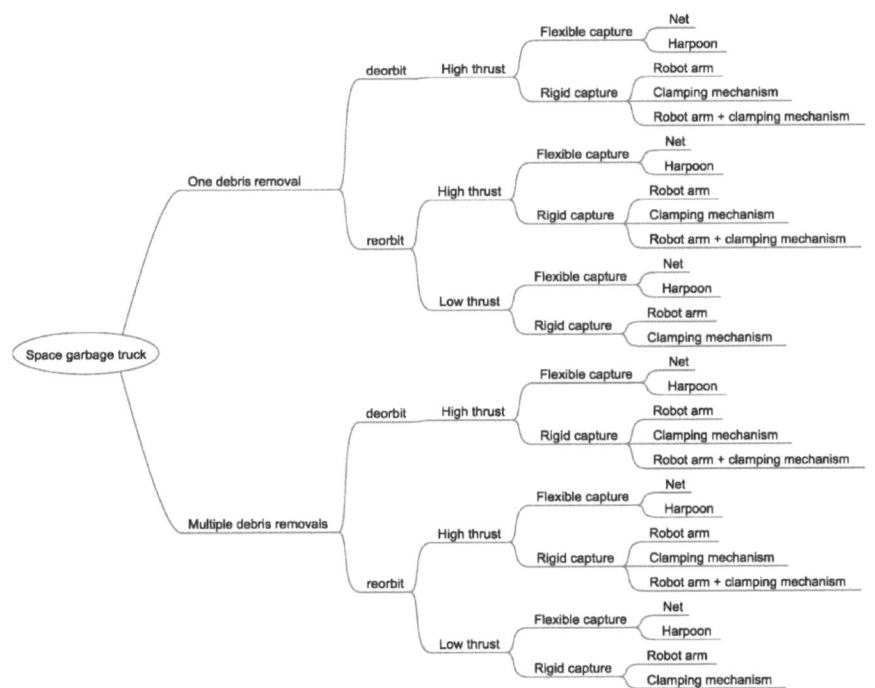

Figure 54: Space garbage truck system trade-off table

Let's look in a bit more detail at the table. Starting from the chaser, shown as 'space garbage truck', there are two branches possible: the objective to have one debris removal, and the objective to have multiple debris removals. For the case of one debris removal, we can deorbit the debris into the atmosphere or re-orbit the debris to an altitude above 2000 km. For the deorbit case, the service module will need to be based on high thrust in order to achieve an accurate reentry corridor. For the re-orbit case, we have two branches again: a service module based on high thrust and a service module based on low thrust. We then get more branches for the capture technique: for either case of the service module we can apply a flexible or rigid type of capture technique. For flexible techniques we have two new branches where we consider the net and harpoon options. For rigid techniques we have two or three branches: the robot arm and the clamping mechanism options. However if we are reentering very large satellites a robot arm may not be give a firm enough grip to withstand the loads from the very high thrust propulsion system (the larger the debris, the higher thrust force is required) and therefore a combination of robot arm and clamping mechanism may be required. For the case of re-orbiting using low thrust, the propulsion system will give a very low force and a combination of robot arm and clamping mechanism is not required.

For the objective of removing multiple targets, the branches are quite similar, but there are a few differences. If we are to deorbit multiple targets, the service module would be a carrier that just brings several payload modules to several targets. The payload module would be attached to the target and using its own high-thrust propulsion system, would deorbit the target. However the service module could be based on either a low thrust or high thrust propulsion system. If there are no constraints on mission duration, the service module could be based on low thrust while the payload modules are high thrust.

In the next section we will look at some examples studied by various space agencies.

EXAMPLES OF DEBRIS REMOVAL MISSION DESIGNS

ESA has studied a mission called 'e.deorbit' under its Clean Space initiative. The mission objective is to actively remove a large ESA owned space debris from orbit. Several mission options were designed simultaneously, in particular a flexible and a rigid capture technique. Different companies were contracted with each giving their own design. Some companies chose for the net option as preferred flexible solution, others chose for the harpoon. Within the trade-off table of the previous section, we can highlight the chosen branches.

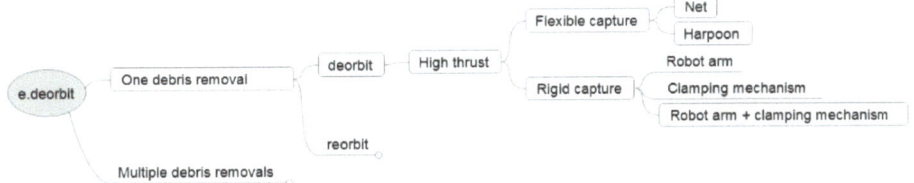

Figure 55: e.deorbit trade-off table. The chosen design options are encircled

The following pictures show some of the e.deorbit designs.

Figure 56: e.deorbit capturing the target with a net. Credits: Airbus Defence and Space [RD15].

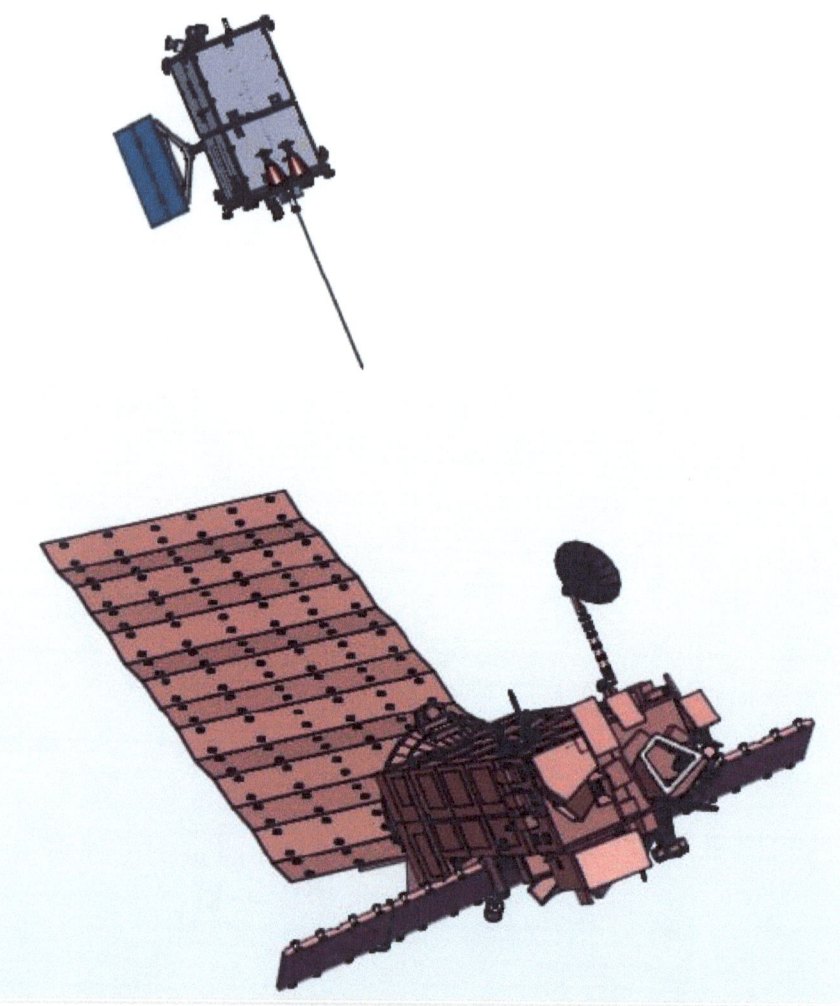

Figure 57: e.deorbit shooting a harpoon at the target. Credits: Thales Alenia Space [RD16].

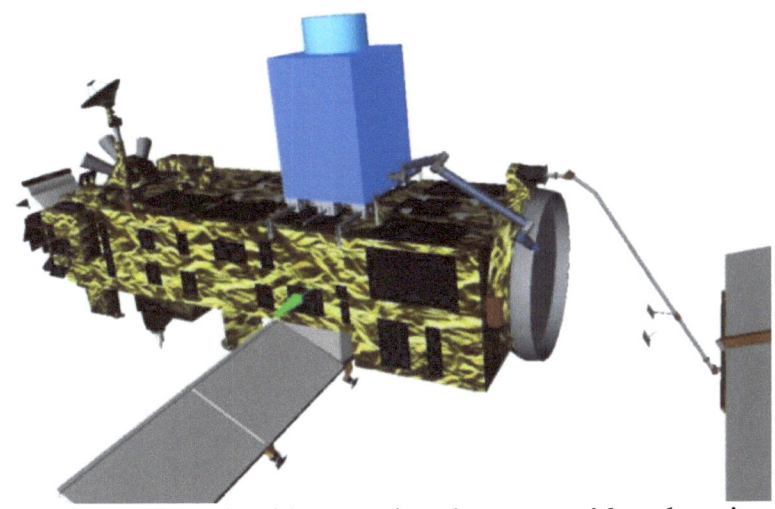

Figure 58: e.deorbit capturing the target with a clamping mechanism and a robot arm to the Launcher Adapter Interface of the target. Credits: OHB System AG [RD17].

The German space agency DLR (Deutsche zentrum für Luft und Raumfahrt) studied a mission called DEOS (Deutsche Orbitale Servicing Mission) which had prime objectives to approach a target, perform servicing operations and to deorbit it. As a technology demonstration mission, DEOS would bring its own target. Therefore both chaser and target were to be launched together, decouple in space and then the mission would continue to reach its objectives (approach, service and deorbit). The servicing objective could only be reached using a robot arm. This would lead to the following trade-off:

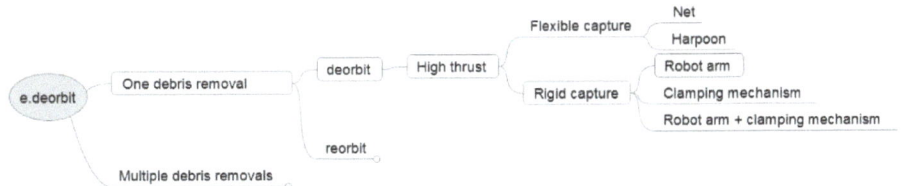

Figure 59: DEOS trade-off table. The chosen design option is encircled

The picture below shows the DEOS chaser on the right, which has captured the target on the left using a robot arm.

Figure 60: DEOS target (left) captured by the chaser (right). Credits: DLR.

The French national agency CNES (Centre National d'Études Spatiales) has also studied ADR in their OTV (Orbital Transfer Vehicle). Though many different options were studied, one option is the removal of multiple objects by means of orbital kids which are installed using a robot arm. This would need to be launched with a heavy rocket such as the Ariane 5. The branches that lead to this selected option are shown below.

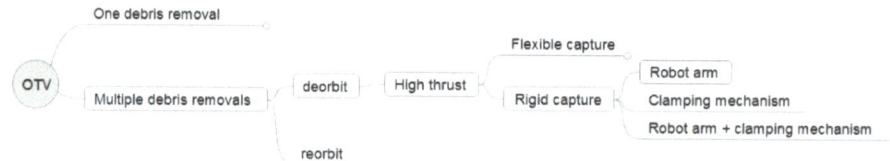

Figure 61: OTV trade-off table. One design option is encircled

Figure 62 below shows a design option of OTV.

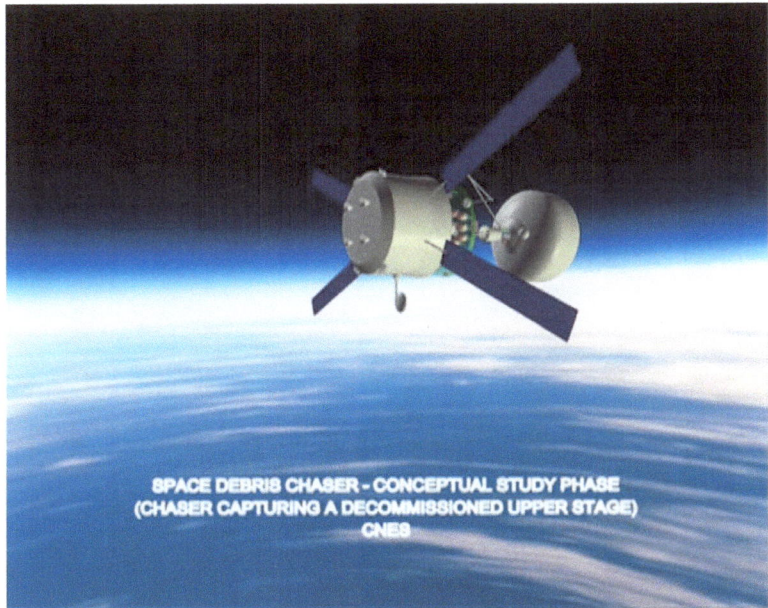

Figure 62:   OTV design. Credits: CNES.

Anno 2015, no ADR mission has gone into implementation phase, meaning to procure equipment and build it. All missions are still in design phase.

CHAPTER 6:   PLANS FOR THE FUTURE

Now we know the current plans for space debris removal, and what it takes to design a space garbage truck, what can we say about the future? Could a space garbage truck be the solution and how viable is it to produce one, or many? And can we dream about the far future?

THE MOVIE 'GRAVITY' AND THE KESSLER SYNDROME

In 2013, Warner Bros Pictures released the science fiction movie 'Gravity'. In the movie, a Russian strike on a space debris object causes a cloud of debris similar to what happened in 2007 with the Feng Yun-1C test. The cloud of debris starts hitting other satellites, which in turn creates more pieces of debris and lead to a chain reaction. Within a short time, most of the satellites in LEO are destroyed and the main character in the movie, astronaut Ryan Stone who was just doing a spacewalk close to the Space Shuttle, needs to fight for her life. Within no time, both Space Shuttle as the ISS are destroyed completely by the debris cloud.

Figure 63:   Still from the trailer of the movie 'Gravity', where the entire ISS is destroyed by space debris. © Warner Bros Pictures [RD18].

As one could expect from a movie, the event is dramatized: within one hour all satellites within LEO have become inactive or destroyed due to the chain effect of debris collisions. However, it is clear from the trend of Figure 3 that the debris cloud around Earth does increase, and moreover the possible destruction of space debris and hazard it can cause to human life is well presented in Gravity.

In fact the chain effect <u>does</u> exist and has already started! And the person who predicted this is called Donald Kessler. Kessler, an astrophysicist working for NASA at Houston, studied the impact of meteorites on Gemini spacecraft that had returned to Earth, in 1965-1966. By 1970, he began to consider if pieces of space debris could be caused by satellite collisions, similar to meteoroids being the product of colliding asteroids. Six years later, NASA was considering launching dozens of giant solar power satellites (SPS) to beam solar power to Earth, and Kessler was asked to investigate the environmental effects of such an endeavor. His conclusion was that an SPS break-up caused by collision would create a large number of debris objects. And all of them could cause more collisions. In July that years he gives a warning in a report that fragmentation by impact between debris pieces will exponentially increase the debris population. The following years, he publishes a number of reports describing the debris problem and proposals to trace space debris, and in 1978 he publishes, with colleague Burton Cour-Palais, a research article called 'Collision frequency of artificial satellites: the creation of a debris belt'. In this paper it is predicted, what is now referred to as 'Kessler syndrome', that collisional break-up will become a new source of orbital debris and that the debris flux will continue to increase of time even if no new satellites or rocket bodies are put into orbit around the Earth.

Figure 64:   Donald Kessler. Source: SpaceVision 2014 [RD19].

This is a major conclusion and somewhat frightening, and recent studies have shown that the Kessler Syndrome has indeed started: if we are to stop launching new objects to space now, the amount of space debris will still continue to increase. Things will not evolve as quickly as in the Gravity movie though. Nevertheless a doubling of the current count of objects could occur within the next few decades. If nothing is done, in the next century the Earth would be surrounded by a huge cloud of space debris making access to space impossible.

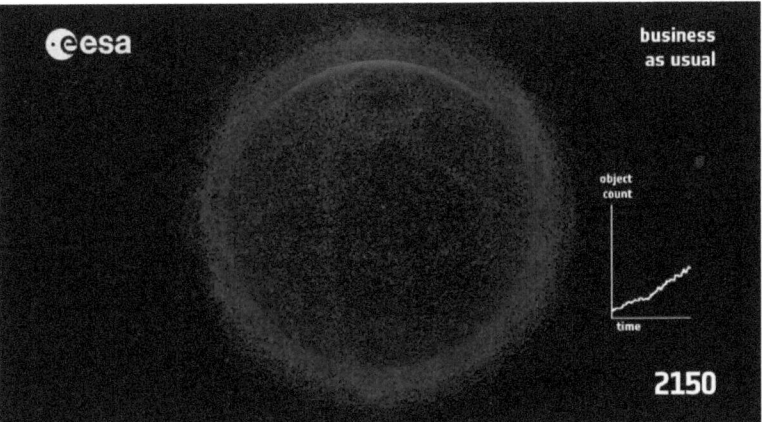

Figure 65: Simulation of space debris in the year 2150 by ESA. Credits: ESA.

The rules set up by the U.N. COPUOS should prevent this unintended increase in space debris and therefore stabilize the graph of Figure 3. However only future space garbage trucks could drastically reduce the amount of space debris by removing large objects with high chance of collisions. Perhaps one of the best features of the Gravity movie is that it made millions of viewers aware of the problem and possible consequences of space debris.

FINANCING THE SPACE DEBRIS REMOVAL BUSINESS

This entire book has been dedicated to why and how we can decrease the number of space debris objects in space, but the million dollar question is still unanswered: who is going to pay for all of this? And why did companies not set up a business, years ago to make money on space debris removal?

National and international space agencies like NASA and ESA could set up a first mission (like ESA's e.deorbit mission) to remove a target and develop all the required technologies for debris removal, so that companies within the participating countries have gained the know-how and could continue to remove debris for future customers. However the question is who those future customers are.

Would it be space agencies, giving orders to clean up their own debris? It is not very likely; space agencies are normally financed by tax payers. For example, ESA consists of many member states (countries), which means that these countries, being stakeholders, have important votes in decisions taken for the agency. For example whether or not the agency should remove many objects in space, rather than invest the money in scientific missions learning about the origins of the universe, or Earth observation missions keeping a close eye on pollution of the stakeholders. Whereas first space removals may come from missions like DEOS or e.deorbit, it will be difficult to convince countries to do multiple follow-up space debris removals in place of scientific missions and other missions that are important to the member states.

Would it be space agencies, financing missions to clean up other nations' space debris? This is even less likely. Why would one nation clean up the garbage of another nation? Remember that ownership remains with the launching state, so one nation cannot just remove debris from another nation because it is 'in the way'. A legal framework would be required that transfers liability from one nation to the other.

Would a government contract a company to remove one of their space debris objects? The question is: why would they? No laws were broken in the past when satellites and rocket stages were left in orbit without properly deorbiting them. The term 'negligence' may play a role in the future for taking decisions to remove space debris. Even if a country was not violating laws by leaving space debris, if that debris objects suddenly starts to hinder a number of other missions of other nations to do their job, or even become hazardous to humans, that country may decide to remove the debris after all.

Space debris removal is a new thing, and therefore trying to insure such a space mission will not be easy. Insurers are typically not very fond of insuring something that has never been done before (meaning, it is a high risk). If they do insure, it is most likely at an outrageous fee and probably not covering the entire investment. Companies wanting to start a business on ADR will therefore need to make large investments at high risk, and will need a number of removal orders in order to gain their investment back and start making a profit. Based on the arguments given above, there will be a very small chance that agencies and governments are willing or capable of placing several orders of space debris removals.

However still, something must be done. We cannot keep polluting our space environment, in the same way that we cannot keep polluting our ground and air environment. As we now see a global movement towards a cleaner ground (reducing CO_2 production of cars for example) and a cleaner air (reducing flight traffic pollution), we need a global movement towards a cleaner space too. Decisions will need to be taken on intergovernmental level and perhaps a new intergovernmental agency is required to handle space debris removal. However how would such an agency be financed? Can we raise 'space debris removal tax' in a similar way that we pay waste disposal tax when we buy large electronic equipment? Perhaps this is the only viable solution.

Taxes are different depending on the pollution created. A car with a high CO_2 emission requires the user to pay more taxes than a car with low CO_2 mission, in many countries. A washing machine with an A label is less polluting that a washing machine with an E label.

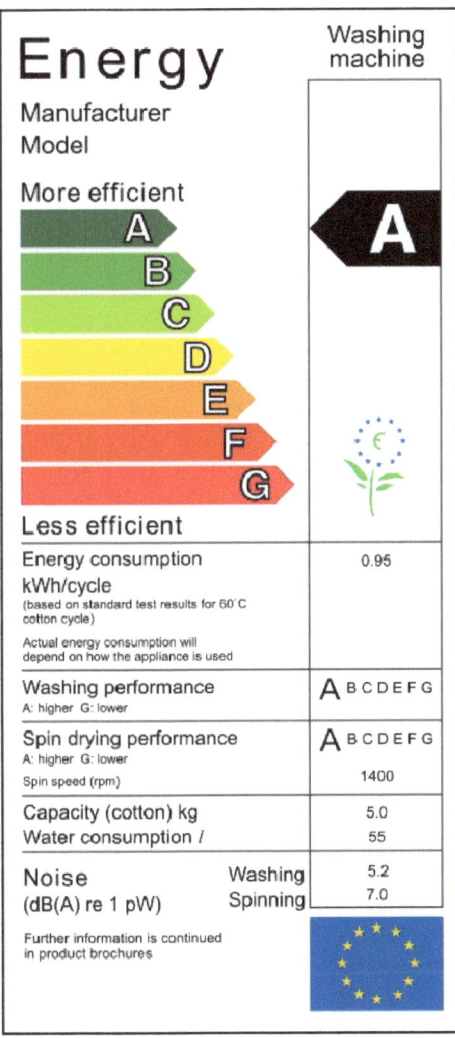

Figure 66: Washing machine energy label. Credits: Chris828 [RD20].

In a similar fashion, satellites in densely polluted orbits, or of high mass, or carrying nuclear energy, could get a bad energy level, whereas small satellites already reentering within 25 years could get a good energy level.

However the life of a satellite mission starts at Earth. So for a true energy label, taking into account pollution on ground, air and space, the entire design and launch process should be considered as well. Satellites that require a 100 people to design and build it, in air conditioned offices over a period of 10 years, spread out over different countries so many team members need to frequently fly using airplanes to meetings, will get a worse label for the ground part. If a satellite only requires ten people to work on it, all on the same site, for only two years, it will get a better label. When the satellite is launched by a launcher with toxic plumes, it will be worse than when the satellite is launched using non-toxic plumes.

Satellite operators, who are launching their satellite, would then need to pay the tax to the intergovernmental organization which would use the income for ADR. Energy labels could then be used by the intergovernmental organization to prioritize which debris needs to be removed first. When a selection is made, contracts could be awarded to space industry around the world to build and launch the removal mission.

REPAIRING SATELLITES OR DEVELOPING NEW SATELLITES IN SPACE

Let's look back at Figure 20 function 5 which is the final function of a garbage truck on Earth: to dispose OR recycle the garbage. Is it possible to recycle garbage? Could this be used to create new satellites or could we even repair satellites to start functioning again?

In order to repair satellites we need to know the cause of malfunctioning. Normally operational satellites require fuel to stay in their exact operational orbit. The fuel on board is limited and at some point in time the satellite will run out of fuel. This often marks the 'end of life' of a space mission. In order to repair a satellite like this, we could consider refueling it. Unfortunately current satellites are not made to be refueled. A repair satellite could approach and capture the dead satellite, stick a refuel line into it and fill it up. Possibly this could be done via the thruster of the satellite as normally the ignited fuel with flow out of here, but valves need to be opened, remaining fuel and pipelines within the satellite may be frozen, membranes may be polluted, and some satellites have different fuel mixtures than others. Other than that, the battery may be depleted and passivated in order to avoid explosions, and operating systems may be wiped out in order to avoid the satellite from accidentally switching on again. All these features will need to be restored when repairing a dead satellite. It will be easier to refuel the satellite when it is still alive, similar to what is done with airplanes. And even more easy if the satellite is built to be refueled.

Figure 67:   Example of a refueling mission. Credits: NASA.

A robot arm would be the obvious choice for a mission like this. Several operations would need to be performed, and most likely more than one robot arm is required to hold the satellite and perform service operations to it.

A propulsion system is often a cause for failure on a space system. Repairing a propulsion system or any other type of system inside the satellite would be difficult as the satellites would need to be opened. NASA has successfully serviced the Hubble Space Telescope, but it required a number of astronauts to do this, as well as dedicated repair tools, and equipment designed specifically for the Hubble. A service satellite carrying 'general equipment to fit most satellites' will not be easy to make as many satellites, in particular in LEO, are custom made. In the GEO ring however, often service module designs are reused for different payload modules, so more standardization is present in that orbit. So while the business case for ADR focuses on LEO, the business case for in-orbit servicing focuses on GEO.

However perhaps we can find a way to mix the two business cases, which would make it easier for companies to make profit. Certainly the two businesses have similar implementations such as performing a close rendezvous with the target and capture it. So a combination could make sense from a future business point of view.

How could the far future look like? Can we dream of a future where we can even recycle satellites in space? As mentioned before, different satellites will require different equipment and different types of fuel. However the same is true for cars on Earth, and while repair workshops could either have all kinds of replacement pieces in stock, the alternative to ordering each piece once it is requested, is to make it on the spot. 3D printing techniques are getting more popular, more affordable, lighter in terms of production machines, and produce stronger products than years ago. If future access to space will get cheaper, could we have future recycling stations where space garbage trucks could bring their debris objects collected as garbage? Space stations large enough could have engineers on board who would dismantle the defunct satellites and build new ones depending on new mission requirements and objectives. Missing equipment could be made in space using 3D printing techniques. The new satellites built in the recycle station do not need to be launched from Earth so will not be subject to the harsh launch environment (in particular in terms of shock levels). They could simply be released from the space recycle station, move to their new orbit and start their new mission. Operational satellites could even pass by the space recycle station to fuel up their tanks and continue their mission.

Figure 68:   Artist impression of a future space recycle station with several space garbage trucks arriving. Credits: MaDe BV.

How far or how near is this future? With the current expensive access to space, and no debris removal or robotic servicing mission ever launched, this future is not very near. However the threat of space debris is ever increasing operational satellites today, so we need to start acting today. Nations around the world have already demonstrated successful approaches of one satellite to another satellite, even uncooperative ones. And as ESA is designing their first ADR mission, as NASA has successfully demonstrated robotic refueling on ground, and as the first 3D printer was just launched to the ISS, this future is getting near soon.

REFERENCES AND RECOMMENDED WEBSITES

REFERENCES

RD1.   Mika   McKinnon,   *A   History   of   Garbage   in   Space*, http://space.io9.com/a-history-of-garbage-in-space-1572783046, 5/07/20014

RD2.   The NASA Orbital Debris Program Office, *Orbital Debris Quarterly News*, Volume 11, Issue 2, April 2007.

RD3.   Wikipedia page, *Skylab*, http://en.wikipedia.org/wiki/Skylab, 22 January 2015

RD4.   Celestrak webpage, http://www.celestrak.com

RD5.   David S. F. Portree and Joseph P. Loftus, *Orbital Debris: A Chronology*, NASA/TP-1999-208856, January 1999.

RD6.   Wikipedia   page,   *Garbage   Truck*, http://en.wikipedia.org/wiki/Garbage_truck, 25 December 2014.

RD7.   The NASA Orbital Debris Program Office, *Orbital Debris Quarterly News*, Volume 18, Issue 2, April 2014.

RD8.   Plot created using the Systems Tool Kit (STK), Developed by Analytical Graphics Inc., www.agi.com

RD9.   The NASA Orbital Debris Program Office, *Orbital Debris Quarterly News*, Volume 19, Issue 1, January 2015.

RD10.   U.N. Department of Economic and Social Affairs, *World Population Prospects: the   2012   Revision*, http://esa.un.org/wpp/unpp/panel_population.htm.

RD11.   Zach   Wilson,   *A   Study   of   Orbital   Debris*, http://ccar.colorado.edu/asen5050/projects/projects_2003/wilson/, December 18, 2003.

RD12.   Maps from US Department of State Geographer, US Geological Survey. Data Source: SIO, NOAA, US Navy, NGA, GEBCO

RD13.   Max Planck Institute für Sonnensystemforschung, *OSIRIS – The Scientific Imaging System for Rosetta*, https://www2.mps.mpg.de/de/projekte/rosetta/osiris/index_print.html, 04-09-2013

RD14.   Arianespace website, www.arianespace.com

RD15.   Stephane Estable, *e.deorbit Symposium presentation*, e.deorbit Symposium, The Netherlands, 6 May 2014

RD16.   Carole Billot, *e.deorbit Symposium presentation*, e.deorbit Symposium, The Netherlands, 6 May 2014

RD17.   Richard Haarmann, *Baseline Concepts for the Kayser-Threde Team*, e.deorbit Symposium, The Netherlands, 6 May 2014

RD18.   The Register, Honey, *I BLEW UP the International SPACE STATION*, http://www.theregister.co.uk/2013/05/17/gravity_bluearc_averre/, 17 May 2013.

RD19.   SEDS-USA, *SpaceVision 2014 Key Note Speakers*, https://spacevision.seds.org/speakers/, 2014.

RD20.   Wikipedia page, *European Union energy label*, http://en.wikipedia.org/wiki/European_Union_energy_label, 17 September 2014.

RECOMMENDED WEBSITES

Clean Space: www.esa.int/cleanspace

DEOS website: http://www.dlr.de/rd/en/desktopdefault.aspx/tabid-2266/3398_read-36724/

e.deorbit 2014 Symposium presentations: https://indico.esa.int/indico/event/46/material/slides/

ESA Space Debris website: http://www.esa.int/Our_Activities/Operations/Space_Debris

IADC website: http://www.iadc-online.org/

JAXA electrodynamic tether: http://www.aero.jaxa.jp/eng/publication/magazine/sora/2004_no02/ss2004no02_02.html

NASA Orbital Debris Program Office: http://orbitaldebris.jsc.nasa.gov/

NASA Orbital Debris Quarterly News: http://orbitaldebris.jsc.nasa.gov/newsletter/newsletter.html

OTV website: http://www.cnes.fr/web/CNES-fr/8617-gp-quel-chasseur-de-debris-pour-demain-.php

U.N. Outer Space Treaty: http://legal.un.org/avl/ha/tos/tos.html

UN COPUOS: http://www.unoosa.org/oosa/COPUOS/copuos.html

TABLE OF ABBREVIATIONS

ACRIMSAT	Active Cavity Radiometer Irradiance Monitor Satellite
ADR	Active Debris Removal
AOCS	referred to as Attitude and Orbit Control System
ATV	Automated Transfer Vehicle
CAM	Collision Avoidance Maneuver
CNES	Centre National d'Études Spatiales
COPUOS	Committee on the Peaceful Uses of Outer Space
DEOS	Deutsche Orbitale Servicing Mission
DHS	Data Handling System
DLR	Deutsche zentrum für Luft und Raumfahrt
ESA	European Space Agency
FSOA	French space operations act
GEO	Geostationary Orbit
GOCE	Gravity Field and Steady-State Ocean Circulation Explorer
HAPS	Hydrazine Auxiliary Propulsion System
IADC	Inter-Agency Space Debris Coordination Committee
IDEF	Integrated Definition
ISS	International Space Station
ITU	International Telecommunications Union
LADEE	Lunar Atmosphere and Dust Environment Explorer
LDEF	Long Duration Exposure Facility
LEO	Low Earth Orbit
LIDAR	Light Detection And Radar
LRO	Lunar Reconnaissance Orbiter
NAC	Narrow Angle Camera
OSIRIS	Optical, Spectroscopic, and Infrared Remote Imaging System
OTV	Orbital Transfer Vehicle

SPOUA	South Pacific Ocean Uninhabited Area
SPS	solar power satellites
SSN	Space Surveillance Network
SSO	Sun-synchronous orbits'
TC	Telecommand
TM	Telemetry
TPS	Thermal Protection System
TSS	Tethered Satellite System
TT&C	Telemetry, tracking, and command
U.N.	United Nations
WAC	Wide Angle Camera